THE COLLECTIONS AND PROGRAMS
OF THE
AMERICAN ANTIQUARIAN SOCIETY

The Collections and Programs of the American Antiquarian Society

A 175th-Anniversary Guide

BY THE STAFF OF THE SOCIETY
WITH A FOREWORD BY
JILL KER CONWAY
AND PHOTOGRAPHS BY
STEPHEN DIRADO

WORCESTER
AMERICAN ANTIQUARIAN SOCIETY
1987

The publication of this book was made possible
by a generous grant from
The William and Flora Hewlett Foundation
made in honor of the Society's 175th anniversary.

LIBRARY OF CONGRESS CATALOGING-IN-PUBLICATION DATA

American Antiquarian Society.
The collections and programs
of the American Antiquarian Society.

Includes index.

1. American Antiquarian Society. I. Title.

E172.A3642 1987 973'.06 87-14480

ISBN 0-912296-93-3

PRINTED BY MERIDEN-STINEHOUR PRESS, LUNENBURG, VERMONT

Table of Contents

Foreword

I STILL recall the astonishment I felt when, as a graduate student in history, recently arrived from Australia, I made my first trip to the American Antiquarian Society. Although that first journey from Cambridge to Worcester took place twenty-six years ago, my memory of it remains vivid because nothing in my previous scholarly life had led me to anticipate that I would find in the town of Worcester a research library of such grandeur, created and sustained by the efforts of private individuals.

Working in my native land as a young historian, I had known about such scholarly tools as Evans's bibliography of American imprints and Clifford Shipton's monumental study of Harvard graduates. What astonished me was to find that the learned society that had produced these aids to scholarship had also accumulated the great collection of materials on the history of colonial and nineteenth-century America on which they were based, and that it was situated in a town of modest size, rather than being located in a great metropolitan center. A day spent exploring the Society's collections only served to increase my wonder. In the Council Room one could gaze on the Mather Family Library. The stacks would yield the actual pamphlets, broadsides, and newspapers of the Revolutionary era. There seemed to be no town, canal, or western scheme of settlement that was not amply documented in the Society's rich and varied collection. And yet this entire archive was the creation of private individuals, inspired by a belief in the importance of history and determined that the records that document the colonial experience and the building of the American nation during the nineteenth century should be collected, preserved and made available to scholars.

All students of American society read Tocqueville's comments on the importance of voluntary associations in the shaping of American culture. The examples of such associations usually given in the history texts are the hospitals for paupers or the self-improvement societies founded by such practical reformers as Benjamin Franklin. For the humanist the vitality and breadth of support committed by a nation of three million people to voluntary associations for learning such as the American Academy of Arts and Sciences and the American Antiquarian Society provide the most dazzling instances of voluntarism in American culture. This patriotic commitment to learning given by individuals in their capacity as private citizens has made the shape of American culture different from all other New World societies because of the private institutions of learning it has nourished. Schools, colleges, universities, and learned societies have all flourished in this country through their capacity to attract private support. As a result, intellectual life in the United States is free from the bureaucratic constraints that affect education and learning in most modern democracies.

It is common for teachers of history in other parts of the English-speaking world to bewail the fact that their students find the study of American history more colorful and more interesting than their own history. While this must in part be understood as a consequence of American economic, political, and military influence in the twentieth century, the fascination exerted by American studies in most societies today is due in part to the tradition of historical writing nourished by collections of the scope built by the American Antiquarian Society and served by the expertise and dedication of scholarly staffs whose mission in life is to make the collections available in ways that serve the needs of all those who seek to learn at first hand about the American past. Because it is not the creation of the state or of a particular profession or a particular cultural bias, but rather exists to build collections that illuminate American life, the Antiquarian Society has amassed a collection of materials about the life of the American people that is unrivaled for its richness; one that anticipated the modern concern with social history and the lives of ordinary people.

The founder of the American Antiquarian Society, Isaiah Thomas, was a printer deeply aware of the political issues of his day. Thus, the collection has, from its inception, preserved the records of the press in American society, and it presents the researcher with a record of American print media that makes the aspiration to comprehensive research genuinely possible.

Today, one hundred and seventy-five years after Isaiah Thomas began to recruit friends to support his plan for preserving the records of the Revolution and the events that led up to it, the library of the American Antiquarian Society is one of the major partners in the Research Libraries Group. RLG, a consortium of the nation's leading research libraries, coordinates cooperative projects that improve nationwide access to research materials. In the field of library science the Society is one of the leaders in the development of online cataloguing for rare and other special materials, and its skills in conservation enable it to serve a broad constituency concerned with the preservation of rare books and print materials.

Since 1970 the Antiquarian Society has played an increasingly important role in fostering scholarly research in American history and culture. Its fellowships support scholars who need to be in residence in Worcester to carry out lengthy research projects, and a generous bequest now makes it possible for the Society to offer residential facilities for its fellows. Its seminars on major research themes bring together faculty and graduate students from academic institutions throughout the country, and its special projects such as the current Program in the History of the Book in American Culture focus attention on important aspects of the American past. Through the sponsorship of the American Antiquarian Society, scholarly editions of significant documents are produced and disseminated to a broader scholarly public, and the Society's program of lectures and conferences serves as a continuing stimulus for interest in American culture. In a less formal sense the Society functions as a sponsor and mentor to scholars, bibliophiles, and antiquarians who are themselves important collectors of Americana. Novices are encouraged, experts are egged on, and burgeoning interests are nourished through the Society's meetings and its publications. The scholarly standards it sets are models for the private collector and amateur historian. The informed public it creates provides the professional historian with a discerning audience.

Representing, as it does, a tradition of humanistic learning unbroken since the Revolution, the American Antiquarian Society is an exemplar of the high cultural standards that the Founding Fathers dreamed could find a place within a society that had turned its back on the aristocratic values of European high culture. This guide to the collections has been one of the most satisfying projects undertaken to mark the one hundred and seventy-fifth year of the Society's operation. Its founding, twenty-five years after the adoption of the Constitution, represents the cultural aspiration to which the Constitution gave political expression. [J.K.C.]

Introduction

THE STAFF of the American Antiquarian Society have chosen to prepare this volume for researchers as our contribution to the Society's 175th anniversary celebration. AAS is not a library whose directors and staff have been anonymous and self-effacing types working quietly to accumulate library resources. From the beginning with Isaiah Thomas to the present under Marcus A. McCorison, our staff members have been noted as strong advocates of and important contributors to the study of American history and bibliography. Also, AAS staff members know the collections thoroughly and work continually to understand the pertinence of that material to new lines of historical inquiry; hence they often become active participants in the scholarly research of our readers.

As we hope this book will make clear, we regard AAS more as a research library than a rare book repository. We do not acquire materials solely for their prestige or bibliographical points, but for their historical importance and their potential usefulness for scholarly research. The Society's librarians have long been known for their prescience in acquiring certain genres, such as newspapers, almanacs, children's books, and women's diaries, long before they became 'legitimate' or popular fields of collecting. In this vein, they have often demonstrated a preference for the homely broadside or pamphlet over a bibliographically elegant work by some literary lion. For this reason, the AAS collections are remarkably rich in the stuff of social history, the story of ordinary people in that experiment called America. Our collections are part of our common heritage. We do not 'own' them the way an individual owns a book or a painting; we hold them in trust and accept the responsibility to preserve and then conserve and make them available to successive generations of scholars.

In preparing this guide, we follow in the tradition of two previous books by Society staff members: R. W. G. Vail's *A Guide to the Resources of the American Antiquarian Society* (Worcester, 1937) and Clarence S. Brigham's *Fifty Years of Collecting Americana for the Library of the American Antiquarian Society* (Worcester, 1958). Vail's book is interesting as a record of its time, but it is no longer useful to today's readers. Brigham's is a memorial to the growth of the library in the first half of the twentieth century and to his own legendary genius for collecting, but it also is no longer pertinent to library patrons because there have been many significant changes in our collections and in their arrangement over the past thirty years.

We designed this book to serve two purposes. First, as a guide for our readers, it seeks to illuminate the rich variety and depth of our holdings and to explain how readers may gain access to them. Second, the accompanying photographs make it a visual record of the Society as it stands in its 175th year: our collections, our physical plant, and our programs. As a repository for the study of history, we are committed as well to preserving our own history.

The articles in this guidebook describing the Society's collections are divided into two main sections, General Collections and Topical Collections. Wherever possible, collections of related content or related scholarly application are grouped together. We urge all researchers to be sure to read the section 'Catalogues and Arrangement of Collections.' The index at the end of the volume provides cross-references from articles. The initials within brackets identify the authors of the articles or groups of articles. The key to their identity lies in the list of contributors at the end of the volume.

The editors wish to thank all the staff members who volunteered to contribute essays. We are grateful to the Society's assistant editor, Sheila McAvey, for her care and diligence in seeing this volume through the press. C. Freeman Keith, of Meriden-

Stinehour Press, brought his usual high standards to bear on the design and production of this book. The Press's president, Roderick D. Stinehour, a vice-president of AAS, generously made this a more handsome volume than it would otherwise have been. Thanks also are owed to Stephen DiRado, who set up his 4 x 5 view camera in all the corners and aisles of the Society to record the visual images you see here. Finally, we gratefully acknowledge the support of Judy and Tony King in funding the photography.

[N.H.B. and J.B.H.]

A Brief History of the Society

THE HISTORY of the American Antiquarian Society is largely the story of the development of its collections and programs, the description of which is the purpose of this book. Accordingly, an extended narrative is not required here. However, there are some benchmarks in the long career of the Society that will be useful to note in this 175th anniversary publication and that will help provide a framework for the following chapters on our collections and programs.

The Society was incorporated by act of the General Court of the Commonwealth of Massachusetts on October 24, 1812, in response to a petition from Isaiah Thomas and his colleagues that they might establish an organization to 'encourage the collection and preservation of the Antiquities of our country, and of curious and valuable productions in Art and Nature [that] have a tendency to enlarge the sphere of human knowledge.' Further, they wished to promote the use of such collections in order to 'aid the progress of science, to perpetuate the history of moral and political events, and to improve and interest posterity.' These objectives still form the basic rationale of our work, which is to collect, organize, and preserve the records of the lives and activities of people who have inhabited this continent. We do so to encourage the study and understanding of the past. Such an activity is valuable to mankind on both secular and religious grounds. We accept that an appreciation of the works of people who came before us is a vital part of a full life. Historical perspective on our lives and times is acquired by a knowledge of the past, thereby giving us a sense of proportion that can be instructive and humbling, hubris being endemic amongst us. As stewards of this portion of the universe, we are bound to treasure God's gifts to us and to pass them on undiminished. Those gifts include a knowledge of the past, an appreciation of the present, and an obligation to the future. In 1813, Thomas instructed his fellow members of the Society in much the same manner: 'We cannot obtain a knowledge of those, who are to come after us, nor are we certain what will be the events of future times; as it is in our power, so it should be our duty, to bestow on posterity that, which they cannot give to us, but which they may enlarge and improve, and transmit to those who shall succeed them. It is but paying a debt we owe to our forefathers.'

The Society was established at a time when all historians were 'amateurs' of history. The members of AAS through most of its first century were the sort of amateur gentleman-scholar that Thomas was. As scholarship became professionalized, AAS took on that shape as well. Thus, through much of its history, AAS has been the sort of institution where both the amateurs and professors of history have collaborated and contributed to the gathering and dissemination of historical knowledge. That ideal endures, although these days the task of providing a forum for exchange between the two groups has become more difficult. Nevertheless, the Society still attempts to provide resources, services, and programs that make the history of our nation interesting, pleasurable, and meaningful to as broad a constituency as possible.

The course of the Society was set by the founder, a printer and publisher, a book collector, and the historian of his trade. He gave to the Society his collection of approximately 8,000 books, some of which, such as an incunable or two, had little to do with America. His other benefactions to AAS included the first library building, erected in 1820, and the residuum of his estate, which supported the Society for some years after his death in 1831. Beyond these gifts, however, was the intellectual vigor with which he endowed the Society. He sought out members from every part of the Union who were expected to send printed and archaeological objects for preservation in the library or in the museum. Thus, in 1819, AAS received the Reverend William Bentley's bequest of his manuscripts, books on American history, and prints; Gov. DeWitt Clinton of New York gave the leaden monument buried at the mouth of the Muskingum River in Ohio in 1749 to mark the claims of the French king to the territories of the Ohio Valley; and Charles Wilkins in 1816 sent to Worcester the mummified remains of an Indian woman that were found in a cave in Kentucky. The earliest historical concerns of the Society centered on the evidences of prehistoric life in North America. Thomas painstakingly edited the disjointed thoughts and jumbled manuscript of Caleb Atwater of Centerville, Ohio, describing the Indian mounds in central Ohio for the first volume of our *Transactions* (1820). It was not until midcentury, after our then-librarian, Samuel Foster Haven, had published the *Archaeology of the United States*, a volume in the Smithsonian's series 'Contributions to Knowledge,' that AAS got out of the ethnological business. However, it should be remembered that one of our original charges was to take an interest in the history of the entire Western hemisphere. Effective involvement on such a scale was far beyond the powers of the Society, although a galaxy of prominent Latin American historians and diplomats graced our membership rolls until well into the twentieth century. The charge lingered as a residual memory and its revival was signaled in 1868 when Isaac and Edward Davis established an endowed book fund, the income from which was designated for the purchase of materials on North America, south of the United States. The interests of Stephen Salisbury III in art (which led to the establishment of the Worcester Art Museum in 1896) and in Latin American archaeology prompted him to sponsor the archaeological expeditions to the Yucatan of Augustus LePlongeon and Edward H. Thompson. Haven lent his full support to these enterprises, and their reports were published in the *Proceedings* during the last quarter of the century.

Following Thomas's death, leadership passed to two successors whose contributions are still evident at the Society. The earlier was Christopher Columbus Baldwin, a high-spirited young man who was rusticated by Harvard College authorities for certain high jinks and who, after brief careers as lawyer and journalist, put his considerable intellect, unbridled enthusiasm, and prodigious energy to the benefit of AAS as librarian from 1827 to 1830 and again from 1832 until 1835. He compiled the several-hundred-page catalogue of the library collections that was issued in 1837. Baldwin was a true bibliophile and antiquary, as he described in his letter to Edward Tuckerman when commenting on the forthcoming marriage of their mutual friend Stephen Salisbury II: 'I have instructed myself in the pedigree of his proposed wife, and I can find no fault with it. This is all an Antiquary may lawfully do with young ladies. I am clearly of the opinion that an Antiquary should not pester himself with a wife: he should do nothing that may diminish his affections with venerable books. You know that we cannot serve two masters, much less two mistresses; and my mistress is my profession, for which I have the most solid affection.'

Baldwin's greatest coup was the securing of the books and pamphlets belonging to Thomas Walcutt of Boston during five days in August 1834. These treasures, he wrote in his diary, 'were put in ancient trunks, bureaus, and chests, baskets, tea chests and old drawers. . . . The extent of them was altogether beyond my expectations. . . . Every thing was covered with venerable dust, and as I was under a slated roof and the thermometer at ninety-three, I had a pretty hot time of it. . . . The value of the rarities I found, however, soon made me forget the heat, and I have never seen such happy moments . . . . I finished packing my things today and helped load them and start them for Worcester. Their weight was forty four hundred and seventy six pounds! I cannot but think that it is the most valuable collection of the early productions of New England authors in the country.'

His unfortunate and untimely death occurred only a year later in a stagecoach accident near Zanesville, Ohio, where he had gone to investigate Indian mounds. It was a sorry loss, indeed, to the Society, for it robbed us of the devotion of a dedicated acquisitor of Americana when only a handful of other institutions were actively collecting such materials.

Baldwin was succeeded by the aforementioned Samuel Foster Haven in April of 1838. He directed the affairs of the Society for the next forty-three years. A scholarly librarian possessing a genial nature but a less dramatic personality than Baldwin, Haven was an effective leader of the Society until his retirement in 1881. He edited volumes three and four of the *Transactions*, as well as the regularly published *Proceedings*. He did much of the preparation of the second edition of Thomas's *History of Printing in America*. Haven played a minor role in the establishment of the American Library Association in 1876. He oversaw the construction of the second library building in 1854 and the 1876 addition to it and maintained and substantially increased both the library and museum collections.

Haven was ably seconded by Stephen Salisbury II, who served the Society as president from 1854 until 1884. Salisbury was a wealthy Worcester entrepreneur who played an active role in the management of the Society and was generous in its financial support, giving, for example, the funds for the addition to Antiquarian Hall. Also, it was President Salisbury who appointed Mary Robinson Reynolds to the library staff as an aide in 1881 at the time of Mr. Haven's retirement.

However, there is one episode that occurred during Haven's tenure that causes pangs of regret to this very day! John James Audubon came to Worcester bearing a letter of introduction dated September 5, 1840, from George Parkman of Boston, addressed to Haven. Audubon called at the library for the purpose of selling to AAS a set of his *Birds of America*. On Audubon's first visit, Haven was busy and the artist was instructed to return later in the afternoon, which he did. The librarian had departed and Antiquarian Hall was locked up. We have not yet obtained a set of that great book and the present incumbent despairs of ever again having the opportunity of doing so.

Haven was followed by Edmund Mills Barton, who had been the assistant librarian since 1866. He served faithfully until 1908 when he retired from an institution that was choking from a plethora of miscellaneous collections and drifting without a clear sense of mission. However, Barton had kept the Society functioning for twenty-seven years, a not inconsiderable achievement at a time when Salisbury was busy establishing the Worcester Art Museum.

Following the death of President Stephen Salisbury III in 1905 and Barton's retirement, the Society underwent a dramatic transformation that would see the construction in 1909–10 of the third Antiquarian Hall from the proceeds of Mr. Salisbury's bequest and the dismantling of the museum. The leadership of the Society passed to President Waldo Lincoln in 1907 and to the new librarian, Clarence Saunders Brigham, in 1908. Lincoln was a descendant of founders Levi Lincoln (Jefferson's attorney general) and Gov. Levi Lincoln, Jr. President Lincoln took an intense interest in the Society until he retired in 1927, coming in almost daily to attend to our business affairs. It was he who, with J. Franklin Jameson and William MacDonald, led the movement to adopt the stance of a research library, to abandon the museum, and to select Brigham as our librarian. An attorney, Lincoln also was an accomplished book collector and left his American cookbooks to the Society upon his death in 1933. Brigham came to AAS from the Rhode Island Historical Society, where, by the age of thirty, he had already established a reputation as a scholar and bibliographer. Lincoln and Brigham reoriented the Society's activities, reinvigorated our affairs, planned the new building, and reestablished the collecting of early American printed works as our principal goal. With a commitment to Charles Evans to assist in the compilation of his *American Bibliography* and with a brand new library building to fill up, Brigham set about his task in earnest. In one year, for example, he obtained more than 7,000 imprints issued before 1821. He went after newspapers with a vengeance and in 1913 began publishing in the *Proceedings* his state-by-state lists that would finally be published in 1947 as the *History and Bibliography of American Newspapers, 1690–1820*. Brigham invested his life in the Society, for years devoting nearly every waking moment to it and overseeing every detail of its work. He supervised the building of one bookstack wing in 1924 and had much to do with the second in 1950. In 1954 Brigham published *Paul Revere's Engravings*, a beautifully produced book that gives a complete account of the patriot's artistic endeavors. He retired in April 1959, after fifty-one years of service to the Society, as one of the nation's preeminent bookmen. During his administration, Dr. Charles Lemuel Nichols, a talented book collector and bibliographer, was an active president who succeeded his father-in-law in 1927, and Calvin Coolidge served as AAS president from his retirement from the White House in 1929 until his death in 1933. Samuel Eliot Morison presided over our affairs from 1938 until 1952 and was succeeded by yet another great bookman, Thomas Winthrop Streeter. Brigham himself was president from 1954 until 1959, while still serving as director, to which office he had been appointed in 1930.

In 1930 Robert W. G. Vail became librarian. While in office, he completed Sabin's *Bibliotheca Americana* (volumes 22–29), picking up where Wilberforce Eames of the New York Public Library had left off. When Vail left AAS in 1939 to become librarian of the State of New York (and later of the New-York Historical Society), Clifford Kenyon Shipton was made librarian. Shipton had been one of President Morison's Ph.D. students at Harvard where he had become the college archivist. Shipton retained that office throughout his twenty-seven years at AAS, spending each Wednesday in Cambridge and Boston where he attended to his archival duties and to his work at the Massachusetts Historical Society as John Langdon Sibley Editor of the *Graduates of Harvard College*. Shipton, in addition to all his other functions, at night at home wrote volumes four through eighteen of 'Sibley's Harvard Graduates,' covering the years 1690–1771.

Shipton's contributions to the Society were significant and substantial. During his first years here, he reorganized the collections of post-1820 materials, removing them from the old alcove arrangement to a classified system, although the scheme that he devised has proved to be needlessly eccentric. Each day each staff member had to perform a stint of assigning call numbers to a set number of books. He then undertook the completion of Charles

Evans's final volume, from the letter N in 1799 through 1800, which was published in 1955. In the same year he began the filming of the *Early American Imprints, Series I, 1640–1800*, with Albert Boni's Readex Microprint Corporation as the publisher of the microforms. Surely this has been one of the most important projects ever undertaken to provide resources for American historical scholarship. Shipton finished this series, which is based on the Evans bibliography, and without pause plunged into the second series, which is based on the Shaw and Shoemaker lists. He had completed the year 1805 at the time of his retirement. He left AAS in midafternoon on August 31, 1967, carrying his green Harvard law school book bag, without saying goodbye to anyone, leaving a tradition of exceeding diligence, devotion to scholarship, and probity. With typical determination and sense of duty to AAS, Shipton, before his death in 1973, completed an index to the Society's *Proceedings*, which was published in 1978.

The relations between Brigham and Shipton were vexed to say the least. One can hardly imagine two more different personalities. Brigham was a bon vivant, with a reputation in the book world as one who enjoyed earthy pleasures to the utmost and an expert and famous builder of martinis. Shipton was ascetic and a teetotaler. If Brigham was lax in some attitudes, Shipton was rigid. Probably not too long a time passed before Brigham came to regret the understanding that Shipton was to succeed to the directorship. He stayed on, long past his prime, while Shipton doggedly waited him out, terribly frustrated, during his own best years. It was not a happy situation into which the present librarian walked on the first of August 1960.

During the Brigham and Shipton years there were several staff members whose importance to the Society must be recorded here. We have alluded to Mary Robinson Reynolds who held, almost, the entire history of the Society in her own memory when she retired in 1941, at age eighty, having served the Society for sixty years. Emma Forbes Waite came to the Society in 1907 and worked effectively with the graphic arts collections until long past her official retirement. Avis G. Clarke came to AAS from Brown University in 1927 as our first professionally trained cataloguer. Her intelligence and determination established standards for analyzing the scholarly usefulness of historical documents and for recording the progress of American printing that are still followed and that are not surpassed elsewhere. Miss Clarke retired in 1970.

Dorothea Nourse Spear was a highly useful assistant to Mr. Brigham until her death, while still in office, in 1959. She had been on the staff for thirty-five years and had done a good deal of research and editorial work for him. Mary E. Brown, a recent high school graduate, came to the Society in 1937, along with several other excellent people, to work on bibliographical projects that were funded by the federal Works Progress Administration. Mary Brown stayed on until her retirement as head of readers' services in August of 1978. She was tenacious in her pursuit of the right book for the right reader, just as she was in tracking down the lost book that had been mis-shelved. A bright and cheerful woman, she maintained order in the reading room, and she, too, made valuable contributions to the historical enterprise.

In the past generation the Society's services to scholarship and the staff to support them have expanded greatly. The staff has grown from eighteen members to fifty. A fellowship program was established in 1970 and the Program in the History of the Book in American Culture in 1983. Our research, publication, and education programs have been expanded or initiated. An office addition to Antiquarian Hall was built in 1970–71 and the Goddard-Daniels House became part of our facilities in the spring of 1982. Twenty years ago, the endowment was $1,500,000; it is now $16,250,000. These changes have taken place through the solid moral and financial support of the Society's Council, members, friends, and our staff. C. Waller Barrett became president in 1964 and was followed by James Russell Wiggins. Wiggins provided energetic leadership from 1970 until he retired as president in 1977, when John Jeppson was elected to that office. Mr.

Jeppson has been generous with his time and has been very attentive to the welfare of the Society's work and finances during the last decade.

Amongst able staff members who are no longer serving is Elliott B. Knowlton, who came as the Society's first development officer in 1967. He retired from that position in 1977, after successfully assisting Frank L. Harrington and the late Howard B. Jefferson, who were chairmen of our development committee, in raising substantial funds for endowment and for the construction of the new wing. Frederick E. Bauer, Jr., came to AAS in 1970 as successor to Miss Clarke, but his many talents soon drew him away from cataloguing tasks and he retired as associate librarian in 1984. Bauer was a skillful and well-beloved manager of our then rapidly growing staff. William L. Joyce, who directed the cataloguing of our manuscripts, was curator of that collection from 1972 until he left in 1981 to go to the New York Public Library. Joyce was particularly effective in working with our visiting scholars, and he planned well-conceived and imaginative educational activities that reached a variety of levels and kinds of historical interests.

The Society is governed by a Council of twenty-two individuals, who appoint the director of the Society and who advise and consent to Society policies and programs. Officers and Councillors are elected by the members of the Society, who now number nearly 500 persons and who come from many states of the Union. Members are elected by their colleagues in recognition of scholarship, for support of cultural institutions, for manifest interest in bibliographical matters, or for distinction as community or national leaders in humanistic affairs. Membership in the Society bears an aspect of honor, but in fact members hold the primary responsibility for the well-being of the institution. The individuals who have been elected to membership since the founding of the Society through April of 1987 are listed in another 175th anniversary publication, *Members and Officers of the American Antiquarian Society, 1812–1987*. Financial support of the Society's activities is derived from income from endowment, from gifts to the annual operating fund, from grants to aid specific projects, and from miscellaneous sources such as fees and the sale of publications.

The next few years in the life of the Society will be as interesting as the last 175. In order to enhance our abilities to serve our nation through the increase of knowledge of its history, we fully expect to continue to enlarge of our financial resources in order to meet inflationary trends and to improve staff services. An addition to Antiquarian Hall is necessary also. A new bookstack that provides full security from fire and flood must be constructed, while adequate work space for the staff and improved facilities for visiting fellows and readers is needed. The future of the American Antiquarian Society waits for no one. We shall enlarge and improve and transmit the splendid history of our nation to those who shall succeed us by keeping the American Antiquarian Society a vibrant and useful place, continuing to pay that debt to our forefathers. [M.A.McC.]

SCHOLARS IN RESIDENCE
WEEKLY CALENDAR

The Library

COLLECTING POLICIES

THE PURPOSE of the American Antiquarian Society has always been to collect and preserve materials that illustrate and enlarge knowledge of the history and culture of the Western hemisphere. In 1968, the Society's Council reaffirmed practices that had evolved over the years and accepted the proposition that our range of collecting would be limited to the territories that became the United States of America and to the former French and English parts of North America from the period of settlement by Europeans through the year 1876. Within those geographical and chronological limits, the proposal dictated that the Society endeavor to collect the printed materials necessary to support advanced research in all aspects of the American experience in a variety of disciplines. This policy was codified in 1985.

Originally, our mandate was interpreted to include ethnological and archaeological objects and relics of colonial life of the entire Western hemisphere. We still retain some of these objects that are pertinent to our historical interests: one of Isaiah Thomas's printing presses, colonial furniture (including clocks), and portraits by or of early Americans. However, by the end of the nineteenth century, it was clear that the Society could not continue to collect artifacts as well as printed materials, and most of the objects were transferred to more suitable institutions, such as the Peabody Museum at Harvard University, the Worcester Art Museum, and the Smithsonian Institution.

It is our mission to maintain the premier research library through which scholars can interpret the meaning of the American experience through printed and manuscript materials. However, we collect American imprints not only because they are carriers of ideas, but also because they are cultural artifacts and are records of the development of printing throughout the country. At the Society, we are convinced that printing is one of the principal agents through which an indigenous culture was developed. Secondary materials published after 1876 are acquired when they pertain to the American period through 1876.

[N.H.B.]

CATALOGUES AND ARRANGEMENT OF COLLECTIONS

The classification scheme used to arrange the American Antiquarian Society's holdings is unique to this library, and reflects both the growth of the collections and the evolution of techniques in providing access to materials. In cataloguing and shelving its holdings, the Society distinguishes between its early American imprints and other materials. That distinction is reflected in the existence of two alphabetized card catalogues at AAS, the Imprints Catalogue and the General Catalogue.

Most American materials through 1820 are fully catalogued, although a substantial number of federal and state documents from the first two decades of the nineteenth century remain uncatalogued. The greatest portion of AAS holdings before 1821 may be found in the Dated Books, Dated Pamphlets, Almanacs, and Broadsides collections; additional materials are located in the Reserve, First Editions, and Bindings collections, as appropriate. These are not classified collections; that is to say, their arrangement on the shelves is not according to a system based on analysis of the content of the work. The Dated Books, Dated Pamphlets, Bindings, and First Editions collections are arranged by main entry; the Reserve and Broadsides collections are arranged by date of publication; and Almanacs are shelved according to place of publication. Catalogue records of these collections may be found in the Imprints Catalogue. In addition to conventional author, title, and subject access, the Imprints Catalogue provides separate date of publication, place of publication, and printers' files, enabling the researcher to locate imprints of a given period, place, or printer.

Efforts are now under way to extend 'imprints level' cataloguing forward to 1830. Broadsides and most pamphlets for the 1820s are now fully catalogued, while the recataloguing of books proceeds more slowly.

For materials printed after 1820–30, or printed outside the United States, the picture is more complicated still. There exist at the Society many discrete uncatalogued collections, each separately shelved and identified by collection name. Such collections tend to be defined either by the physical format (e.g., broadsides, pamphlets) or by the genre of the material (e.g., canal and railroad reports or trade catalogues). In some cases, checklists serve as finding aids.

For catalogued materials, a classification system unique to AAS, devised by librarian Clifford K. Shipton, is employed. Shipton created a number of classifications (e.g., Genealogy, Local History, Biography, Printing History) tailored to the Society's holdings and collecting interests. The initial letters assigned to these classifications are often derived from the Library of Congress Classification System; the resemblance to the LC scheme ends there. The arrangement of materials within each classification is dictated by the materials classed there. Biographies, for example, are shelved by the subject of the work, then by author; local histories by state, then by county, city, or town; and the classification for works on religion is divided into sections representing the various denominations reflected in the Society's holdings.

The Society's General Catalogue provides access to catalogued materials printed after 1820. (The user of the catalogues must bear in mind, however, that materials from the 1820s are migrating from the General to the Imprints Catalogue, and that access to the Genealogy and First Editions collections is provided in the

Imprints Catalogue.) It has been observed that a history of cataloging might be written from the cards found in the General Catalogue, where handwritten half-cards stand next to (and must be interfiled with) computer-generated cards. Cataloguing rules have changed drastically over the years. At AAS, this means that the access provided by the catalogue is inconsistent; for some works full access by author, title, series, and subject is available, while for others no more than a single, brief, handwritten entry will be found. At the Antiquarian Society, more than at most libraries, then, the assistance of the readers' services staff in interpreting the catalogue is of critical importance to the reader.

[A.N.D.]

CATALOGUING PROGRAMS

The past twenty years have seen dramatic changes in the way materials are catalogued at the American Antiquarian Society. Two decades ago, most post-1820 materials were catalogued on typed half-cards. Today, AAS cataloguers employ four different computer systems in the creation of machine-readable entries. Computer-printed cards filed in the Society's card catalogues are at present the most useful product of these computerized records, but we are laying the foundations for catalogues in other formats, among them COM (Computer Output Microform), CD-ROM (Compact Disc-Read Only Memory) and online catalogues. In the interim, AAS's machine-readable records are available through the Research Libraries Information Network (RLIN), a national computer system that makes detailed information on the Society's holdings available to major research libraries around the country and abroad.

The ongoing cataloguing of newly acquired materials and the gradual cataloguing of previously uncatalogued collections has been substantially assisted in recent years through several cataloguing projects. Such efforts, supported by the National Endowment for the Humanities, the U.S. Department of Education, and by private foundations, are a means of providing improved access to specified categories of materials within the Society's collections.

The first such project, funded by NEH, made possible the creation of detailed collection descriptions, in both a card catalogue and in printed volumes, of the Society's manuscript collections.

A second project, also begun in the early 1970s, was the cataloguing of uncatalogued pamphlet literature of the 1820s. During the course of this effort, AAS cataloguers created the first machine-readable cataloguing records of AAS holdings. Materials catalogued were transferred to the Dated Pamphlets collection, and the catalogue cards produced were filed in the Imprints Catalogue, thus providing 'imprints level' cataloguing for the first time for materials printed after 1820.

In 1978, work began on the cataloguing of the Society's collection of American broadsides printed before 1831. That project was completed in 1985. The Society's collection of American children's books for the period 1821–76 is now being catalogued. From 1982 through 1987, the Society was a participant in the United States Newspapers Program, a cooperative effort for the cataloguing of the nation's newspapers. The Society contributed over 14,000 records to the shared database, available through the Online Computer Library Center, Inc. (OCLC) and RLIN.

The largest project now under way at AAS is the North American Imprints Program (NAIP). Envisioned in its broadest terms, NAIP's goal is to provide detailed bibliographical descriptions of and sophisticated access to materials published in the United States and Canada through 1876. Currently, the program focuses on books, pamphlets, and broadsides (but not newspapers or periodicals) printed before 1801, whether held by AAS or by another institution. The results of this major cataloguing effort will become part of the British Library's Eighteenth-Century Short Title Catalogue (ESTC) and will also be supported as an independent database of some 40,000 records, recording the locations of over 120,000 reported copies. Concurrent with this work, NAIP staff members, with U.S. Department of Education funding, are creating a full catalogue of the Readex Microprint Corporation's *Early American Imprints, First Series (Evans)*, which reproduces in microform most of the titles catalogued by NAIP.

The portion of the NAIP file that describes materials held by AAS is available through RLIN; these materials are also accessible through the Society's Imprints Catalogue. Records describing materials not in AAS collections will be available through RLIN's ESTC Special File in 1989; in the meantime, limited access is available at the Society, both online and through printouts of the file. [A.N.D.]

CONSERVATION

From the beginning, the preservation of the printed and manuscript record of the United States has always been an explicit goal of the American Antiquarian Society. The founders of the Society recognized that, without ensuring the preservation of its collections, the acquisition and cataloguing of those collections would be wasted effort. That they took preservation seriously is evident in the motto they adopted for the Society, which proclaims that the collection 'cannot be harmed by the devastation of war nor the ravages of time.'

The role of the conservation department at AAS is to ensure the continued existence of the Society's collections. The department's activities fall into two broad categories: preservation and conservation. The word 'preservation' includes all activities aimed at extending the life of the physical or intellectual content of the material. 'Conservation,' one aspect of preservation, denotes specific treatment of individual items to restore the artifact itself to a near original condition.

The earliest efforts at preserving the collection were to house it in a substantial building in a geographically protected location. In 1819, Isaiah Thomas contracted for the erection of a brick building to provide protection from fire. When additions to that building were proposed in 1831, his will required that the roof sections be covered with slate or zinc, that the floors be covered with brick or tile, and that communication between the main building and the addition be by means of an iron door.

The present building, built in 1910 and the third the Society has occupied, had air conditioning and humidification equipment installed in the stack area in 1963. Ten years later, the system was expanded and upgraded, the roof was more heavily insulated, and the windows were triple glazed with ultraviolet-filtering plexiglass. These changes were instituted to further stabilize temperature and humidity, reduce harmful radiation from the sun, and add to building security.

Conservation activity in the first 170 years of the Society's existence consisted largely of binding newspapers and periodicals. This was done commercially until 1912, when Clarence Brigham hired an English-trained binder, Horace Phillimore, to establish an in-house bindery. Mr. Phillimore worked in the library's bindery for more than forty years, retiring in 1954. Eventually, increased awareness of the scope of preservation took on new urgency and importance, and the conservation department was created with preservation as its sole responsibility.

Conservation has long had an established place in museums, but it is only since 1966 and the Florence flood that libraries and archives have embraced conservation as an important part of their mission. At the American Antiquarian Society, as in other places, the transition from bindery to conservation laboratory has been a gradual one, and by 1960 the staff had become increasingly concerned with practices that damaged, rather than protected, the research materials. Practice in the bindery shifted from binding periodicals to deacidifying and mending paper, as well as using archivally sound material to repair and house the fragile items in the Society's collection. In 1982, a professional conservator was hired to head the department, and the transition was complete.

The Society currently pursues preservation in a number of ways. Central to the preservation effort is the preservation policy drawn up in 1985. This policy sets forth the principles of preservation at AAS and lists the priorities of the preservation program.

The first priority is control of the environment. The library building and its internal environment represent the first line of defense against damage to the collection. A structurally sound brick building protects the collection from water or fire damage. Since heat, light, and moisture accelerate the chemical deterioration of library materials, these are also controlled.

The second priority of preservation activity at AAS falls under the heading 'collection maintenance.' This includes all the

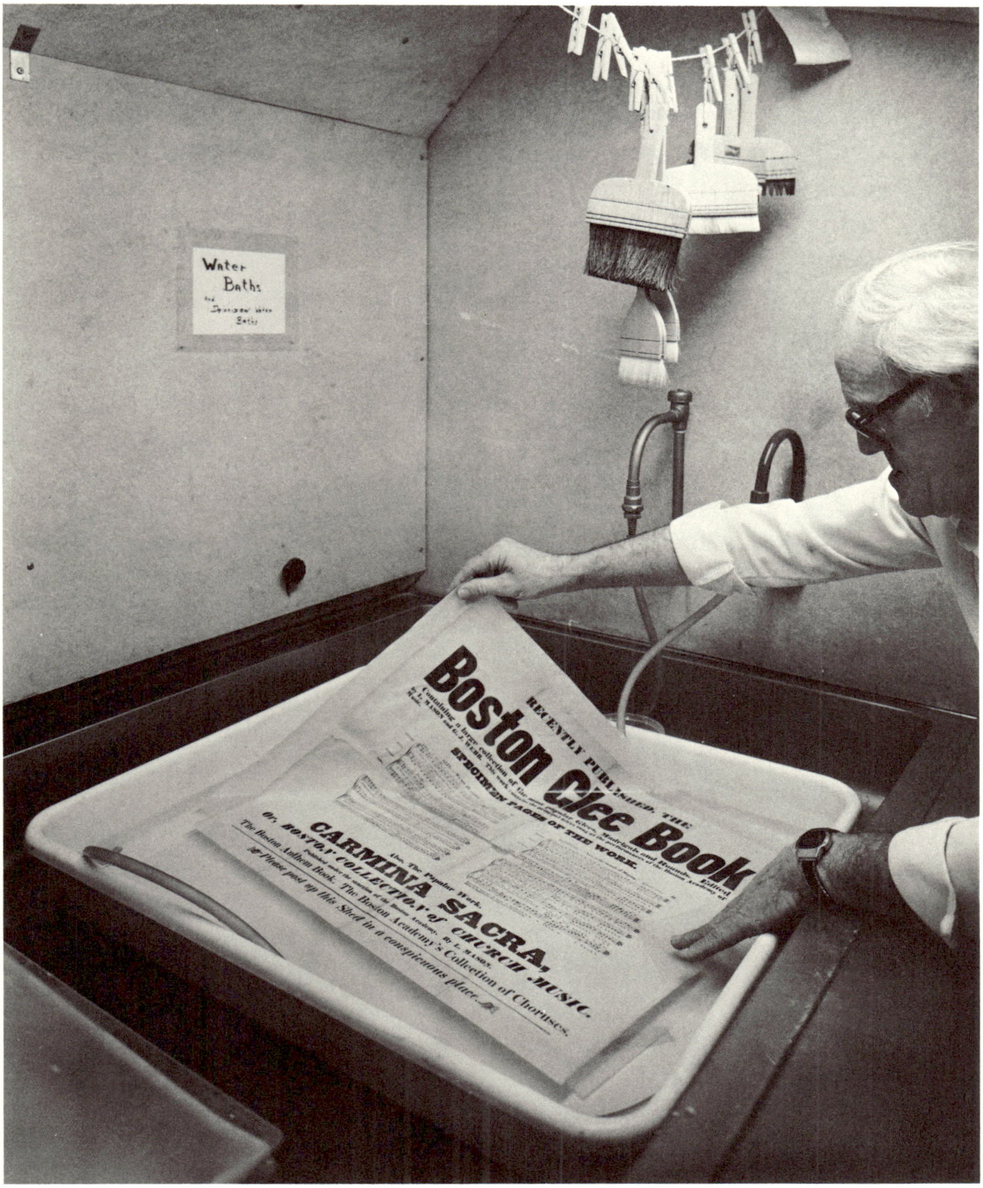
Water Baths
RECENTLY PUBLISHED, THE
Boston Glee Book
SPECIMEN PAGES OF THE WORK.
Also, The Popular Work,
CARMINA SACRA,
Or, BOSTON COLLECTION of CHURCH MUSIC.
The Boston Anthem Book. The Boston Academy's Collection of Choruses.
Please post up this Sheet in a conspicuous place.

policies for handling and storing the material in ways that will best preserve it. Since paper manufacturing processes changed in the 1860s, acidity has been a major source of cellulose degradation. AAS uses special boxes, envelopes, and folders made of acid-free material to house the collection. The Society also has a disaster plan in place that provides for recovery from water damage.

The third priority of preservation is the actual conservation of individual items. Since the collections are valuable to researchers not only for their content but also as artifacts providing evidence of the material culture of their times, the Society strives to preserve both the intellectual content and the physical items themselves. The Society has a fully equipped conservation laboratory capable of handling all necessary treatments, limited only by the size of the sink and drying racks. The conservation treatments follow accepted professional standards. These include reversibility, the use of nondestructive materials, and proper documentation.

Within individual departments, priorities for conservation treatment are established by the library curators and approved by the assistant librarian and the director. Treatment decisions for individual items are made jointly by the appropriate curator and the conservator. Each department needing conservation time is allotted a percentage of the conservation bench time available. Currently, the department has a staff of two full-time employees and a total of 2,440 hours of bench time available per year.

Clarence Brigham noted with pride in *Fifty Years of Collecting Americana* that 'the cost of binding is less than half of what we would have to pay to a commercial bindery.' Although the nature of the work has changed since that was written, in-house conservation treatment still costs only about half of what the Society would have to pay for similar work at a privately run conservation laboratory. [R.C.B.]

Guide to the Collections

GENERAL COLLECTIONS

DATED BOOKS AND DATED PAMPHLETS

WHEN in 1813 Isaiah Thomas enumerated classes of materials suitable for deposit in the fledgling Society's library, he listed *primus inter pares* books, pamphlets, and magazines, 'especially those which were early printed either in South or in North America.' Ever since, Thomas's mandate has been so assiduously carried out that the Society's collection of books and pamphlets 'early printed' in British North America has grown into the largest such collection anywhere in the world. In addition to works printed in the thirteen colonies that became the United States, and in the United States themselves and their territories, the collection includes smaller numbers of Bermudian, Canadian, and British and French West Indian imprints.

The books and pamphlets in the Society's collection touch on every subject of importance in American history, life, and letters through the year 1830, and on every trivial subject as well. The products of big city and small town presses from Abingdon, Maryland, to Zanesville, Ohio, are represented, as are all the ideas for which printers or authors imagined there to be a market, or at least a thirst, in their neighborhood, state, or nation.

The collection is now known by two names, 'Dated Books' and 'Dated Pamphlets,' which reflect discrete shelving arrangements. It comprises approximately 40,000 items that were printed before 1831. This number does not include any almanacs or broadsides, nor does it include *all* of the Society's pre-1831 American books and pamphlets. (Many items printed during the decade 1821–30 are shelved with the Society's general collections and with various uncatalogued collections; other materials are held in the Reserve Collection and in the collections of state and federal documents.)

Until 1974 the Society for practical reasons tended to define 'early' as 'before 1821,' thus following the lead established by the chronological limits of 1821–61 in Orville Roorbach's *Bibliotheca Americana* (1852–61) and the 1639–1820 projected scope of Charles Evans's *American Bibliography* (1903–34). In 1974 the Society redefined 'Dated Books and Dated Pamphlets' to include works printed before 1831. Newly acquired American books and pamphlets printed before that year are now generally shelved in the Dated collections and catalogued in the Imprints Catalogue. As time and resources permit, such items hitherto shelved with the general collections are transferred to Dated Books or Dated Pamphlets and accorded the exhaustive cataloguing long associated with these collections.

From 1927 until 1970, the Society's early American imprints were catalogued by Avis G. Clarke. Miss Clarke's authoritative author, title, subject, and added entry cataloguing for these books was published by the Society in 1971 as *A Dictionary Catalog of American Books Pertaining to the 17th through 19th Centuries: Library of the American Antiquarian Society*, 20 vols. (Westport, Conn.). Miss Clarke also produced files of cards for these books arranged by date of printing, by place of printing, and by name of printer. The printers' file includes cards that synopsize each printer's career; these cards are backed up by 'printer authority cards,' which list the sources, ranging from contemporary newspapers to modern genealogies, for all information given in the printers' file. The cataloguing department continues to add to this file and has expanded coverage to 1831. These files were not published with the *Dictionary Catalogue*, but they are accessible in the reading room to all who use the Society's collections in person. The new machine-readable database of highly detailed bibliographical information on all pre-1801 North American books and pamphlets is available online both at AAS and at all member institutions of the Research Libraries Group through RLIN. For more information on this database, see the sections in this book on catalogues and arrangements of collections and cataloguing programs.

All but a small number of the Society's pre-1820 Dated Books and Dated Pamphlets have been reproduced in microform as the largest part of the Readex Microprint-AAS Early American Imprints series. These two series, Evans 1640–1800 and Shaw-Shoemaker 1801–19, are held by several hundred college, university, and independent research libraries in the United States and fifteen such institutions in Canada and abroad. Cataloguing records for the Shaw-Shoemaker series are available online through RLIN and OCLC; records for the Evans series will be available through RLIN in the near future. Because the Society has made photographic reproductions of the complete texts of its early American imprints available to scholars worldwide, use of the originals is now limited to scholars engaged in specific kinds of research. This policy is part of the Society's commitment to ensure that the fragile originals of all documents in its collections will be available for inspection generations hence. [K.A.]

BROADSIDES

The collection of broadsides at the American Antiquarian Society printed before 1877 is believed to be the most extensive in existence. It is made up of many thousands of items. These single-sheet printed documents were issued locally in response to specific popular or newsworthy events or were otherwise designed for short-lived purposes. Many were subsequently destroyed or put to other uses. Those that survive today provide interesting perspectives on various aspects of the history and culture of the nation.

In 1872 the librarian of the Society, Samuel F. Haven, presented a useful definition of these materials in his semiannual report to the Society. 'Broadsides,' he stated,'are the legitimate representatives of the most ephemeral literature, the least likely to escape destruction, and yet they are the most vivid exhibitions of the manners, arts, and daily life, of communities and nations.

. . .They imply a vast deal more than they literally express, and disclose visions of interior conditions of society such as cannot be found in formal narratives.' The subject matter of the broadsides is remarkably diverse and ranges from the more mundane official government proclamations and regulations, tax bills, and reports of town meetings, to the more interesting contemporary accounts of events in the French and Indian War, or the American Revolution, as well as other unusual occurrences and natural disasters. The collection also contains numerous autobiographies and dying confessions of convicted criminals, theatre playbills, sheet almanacs, publishers' prospectuses, advertisements, newspaper carriers' addresses, patriotic and popular songs, ballads, and poems, broadsides illustrating political party organization and controversy. For the student of nineteenth-century social and cultural history, there is information on a wide variety of local and national organizations and societies that were established to promote industrial and mechanical arts, agriculture, science, public education, religion, the fine arts, and various reform movements. Isaiah Thomas was highly instrumental in preserving many of his own and other printers' most ephemeral pieces, and the Society actively collects broadsides printed before 1877.

Between 1978 and 1985, AAS received two grants from the National Endowment for the Humanities to support the creation of a bibliographically descriptive catalogue, in machine-readable format, of almost 8,000 of the Society's original broadsides and photostats that were printed in the United States from 1639 through 1830. As a result, AAS is able to provide full, scholarly access to its earlier broadsides through complex methods of indexing and information retrieval through RLIN, the shared cataloguing network of the Research Libraries Group. Researchers are now able to retrieve pre-1831 materials through a wide variety of access points, including author, title, multiple subject headings, added entries of personal or corporate names, special genre headings (e.g., broadsides, poems, proclamations, prospectuses, playbills, and advertisements), appropriate bibliographic reference numbers, provenance data, illustration technique, printer, and

It seems he died about the time that some great Comet appeared in 1680 or 81.

A FUNERAL ELEGY,

DEDICATED TO THE MEMORY OF HIS WORTHY FRIEND,

The Learned & Religious

Mr. John Foſter;

Who Deceaſed in Dorcheſter, *the* 9th. *of September*, 1681.

AMongſt the *Mourners* that are met
(For Payment of their laſt love-debt
Unto the dead) to ſolemnize,
With Sighs and Tears, his Obſequies,
Loves Laws command, that I appear,
And drop a kindly friendly Tear.
I'll venture to bewail his *Herſe*,
Though in a homely Country verſe;
Sith to omit the ſame, it were
A crime at leaſt *Piacular*.
Our woful loſs for to ſet forth,
By ſetting forth the matchleſs worth
Of the *Deceaſed*, is too high
For my poor *Rural* Poetry:
And greater skill it doth require,
Than whereunto I may aſpire.
Records declare, how he excell'd
In *Parentage* unparallel'd.
Whoſe Grace and Virtues very great,
He did himſelf *Impropriate*
Unto *Himſelf*; improv'd, withal,
By Learning *Academical*.
His curious works had you but ſeen,
You would have thought Him to have been,
By ſome ſtrange *Metempſychoſis*
A new reviv'd *Archimedes*;
At leaſt you would have judg'd that He
A rare *Apelles* would ſoon be.
Adde to thoſe things I have been hinting,
His skill in that rare Art of *PRINTING*:
His accurate *Geography*,
And *Aſtronomick Poetry*;
And you will ſay, 'twere pitty He
Should dy, without an *Elegie*.
His piercing *Aſtronomick EYE*
Could penetrate the cloudy sky,
And ſoar aloft, ith' higheſt *Sphere*,
Deſcrying *Stars* that diſappear
To common eyes: But *Faith* and *Hope*
His all-excelling *Teleſcope*,
Did help his heaven-born ſoul to pry
Beyond the Starry *Canopy*.
His excellencies here, we find
Were crowned with an humble mind.
Thus (Grace obtein'd and Art acquir'd,
And *thirty three* years near expir'd)
He that here liv'd belov'd, contented,
Now dies bewail'd, and much lamented.
Who knows the *Skill*, which to our loſſe
This *Grave* doth now alone ingroſſe?
Ah who can tell *JOHN FOSTER'S* worth?
Whoſe *Anagram* is, *I SHONE FORTH*,
Preſaged was his Apoge,
By a preceding Prodigie;
Heav'ns blazing Sword was brandiſhed,
By *Heav'ns* inraged wrath, we dread;
Which ſtruck us with amazing fear
Some fixed Star would diſappear:
Th'appearance was not long adjourn'd
Before our Fear to ſorrow turn'd.
Oh Fatal Star (whoſe fearful flame
A fiery Chariot became,
Whereby our *Phœnix* did aſcend,)
Thou art our Foe, although his Friend.
That rare Society, which forth
Hath ſent ſuch *Gems* of greateſt worth,
It's *OAKES* and pleaſant *Plants* by death,
Being pluckt up, it languiſheth:
Thus dye our hopes, and *Harvards* glory
Scarce parallel'd in any Story.
That GOD doth thus our choice ones ſlay,
And cunning Artiſts take away,
The Sacred Oracles do ſhew
A dreadful flood of wrath's in view.
Oh then let every one of you
His rare accompliſhments that knew
Now weep! weep ye of *Harvard Hall*,
With bittereſt Tears! ſo weep We all;
Chiefly let ſuch as were alone
Fleſh of his Fleſh, Bone of his Bone,
Lament indeed, and fill the skyes
With th'eccho's of their doleful cryes.
Let *JAMES*, and let *ELISHA* too, (go,
With *COMFORT*, *STANDFAST*, weeping
THANKFUL, *PATIENCE*, *MARY*
Like loving *Siſters* ſolemnize (likewiſe,
With Sigh's, your greateſt loſſe! but yet
Your *Thankful Hope* do not forget,
With perſeverance to fulfil;
Know, your *Elijah's* GOD lives ſtill:
Stand faſt therefore with *Patience*,
Comfort ſhall be your recompence.
And as you yet ſurvive your Brother,
So be like comforts to your Mother,
Who like *Naomi* ſad, is left,
Of Husband, and two Sons bereft.
So *Bitterly* th' *Almighty One*,
Hath to our weeping *Marah* done.
Grieve not too much: the time draws near,
You'll re-injoy Relations dear,
And All together ſhall on high,
With everlaſting Melody,
And perfect peace *His* praiſes ſing,
Who through all troubles did You bring.

Thomas Tileſtone.

date and place of publication. Access is also available through a chronologically arranged shelflist of the collection. Both of these tools are very efficient means for locating specific items. The elimination of indiscriminate, manual searching for these items greatly enhances the Society's efforts to preserve these extremely fragile, and often unique, research materials.

The broadsides issued after 1830 are not catalogued. Their arrangement and access is either by date of publication or, when appropriate, according to specific genre. The vast majority of broadsides pertaining to Worcester forms a separate collection that is arranged by subject. [C.R.K.]

ALMANACS

The almanac has been called the one universal book of modern literature. In early America it was the most abundant and most indispensable of all publications, a necessity to farmers, navigators, householders, townspeople, the gentry, the professional class, and even to scholars. The almanac had an essential place in homes where no other form of literature entered and where, often, not even the Bible and the newspaper were found. If the almanac had a comprehensive subject, it was: How to get through life. The otherwise dissociated miscellany it contained was indeed rather like that forming the contents of a person's mind going through life each day. Not only an anthology of daily life but a preview of its entire visible cosmic setting through the coming cycle of months was to be found in this stitched-up pamphlet of soft paper, 'Fly blown and tattered, that above the fire / Devoted smokes and furnishes the fates / And perigees and apogees of moons.'

Today, the almanac remains one of the most generally attractive categories of early publications because of the pleasing variety of matter found within it. Besides the calendar pages, it includes farming advice, medical and domestic recipes, literary excerpts in verse and prose, morally edifying passages and celebrations of rural virtue, political exhortations, useful lists of roads

THE
BALLOON
ALMANAC,

For the Year of our LORD, 1796.

BISSEXTILE, OR LEAPYEAR.

Adapted to the Horizon and Meridian of Philadelphia.

LANCASTER,
PRINTED BY WILLIAM & ROBERT DICKSON,
FOR JOHN WYETH, HARRISBURG.

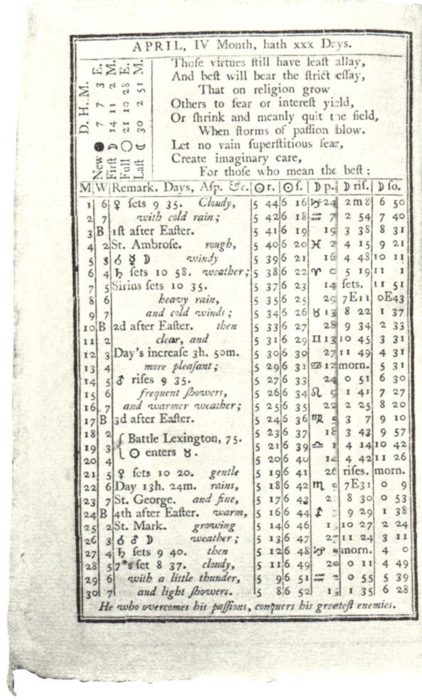

APRIL, IV Month, hath xxx Days.

	D.	H.	M.	
New ●	7	7	3	E.
First ☽	14	11	2	M.
Full ○	21	10	28	E.
Last ☾	30	2	51	M.

Those virtues ſtill have leaſt allay,
And beſt will bear the ſtrict eſſay,
That on religion grow
Others to fear or intereſt yield,
Or ſhrink and meanly quit the field,
When ſtorms of paſſion blow.
Let no vain ſuperſtitious fear,
Create imaginary care,
For thoſe who mean the beſt;

M	W	Remark. Days, Aſp. &c.	☉r.	☉ſ.	☽p.	☽riſ.	☽ſo.
1	6	♀ ſets 9 35. *Cloudy,*	5 44	6 16	♑ 24	2 m 8	6 50
2	7	*with cold rain;*	5 42	6 18	♒ 7	2 54	7 40
3	B	1ſt after Eaſter.	5 41	6 19	19	3 38	8 31
4	2	St. Ambroſe. *rough,*	5 40	6 20	♓ 2	4 15	9 21
5	3	☌ ☿ ☽ *windy*	5 39	6 21	16	4 48	10 11
6	4	♄ ſets 10 58. *weather;*	5 38	6 22	♈ 0	5 19	11 1
7	5	Sirius ſets 10 35.	5 37	6 23	14	ſets.	11 51
8	6	*heavy rain,*	5 35	6 25	29	7 E 11	0 E 43
9	7	*and cold winds;*	5 34	6 26	♉ 13	8 22	1 37
10	B	2d after Eaſter. *then*	5 33	6 27	28	9 34	2 33
11	2	*clear, and*	5 31	6 29	♊ 13	10 45	3 31
12	3	Day's increaſe 3h. 50m.	5 30	6 30	27	11 49	4 31
13	4	*more pleaſant;*	5 29	6 31	♋ 12	morn.	5 31
14	5	♂ riſes 9 35.	5 27	6 33	24	0 51	6 30
15	6	*frequent ſhowers,*	5 26	6 34	♌ 9	1 41	7 27
16	7	*and warmer weather;*	5 25	6 35	22	2 25	8 20
17	B	3d after Eaſter.	5 24	6 36	♍ 5	3 7	9 10
18	2	Battle Lexington, 75.	5 23	6 37	18	3 43	9 57
19	3	☉ enters ♉.	5 21	6 39	♎ 1	4 14	10 42
20	4		5 20	6 40	14	4 42	11 26
21	5	♀ ſets 10 20. *gentle*	5 19	6 41	26	riſes.	morn.
22	6	Day 13h. 24m. *rains,*	5 18	6 42	♏ [illegible]	7 E 31	0 9
23	7	St. George. *and fine,*	5 17	6 43	[illegible]	8 30	0 53
24	B	4th after Eaſter. *warm,*	5 16	6 44	♐ [illegible]	9 29	1 38
25	2	St. Mark. *growing*	5 14	6 46	[illegible]	10 27	2 24
26	3	☌ ♂ ☽ *weather;*	5 13	6 47	[illegible]	11 24	3 11
27	4	♄ ſets 9 40. *then*	5 12	6 48	♑ [illegible]	morn.	4 0
28	5	7*s ſet 8 37. *cloudy,*	5 11	6 49	20	0 11	4 49
29	6	*with a little thunder,*	5 9	6 51	♒ [illegible]	0 55	5 39
30	7	*and light ſhowers.*	5 8	6 52	[illegible]	1 35	6 28

He who overcomes his paſſions, conquers his greateſt enemies.

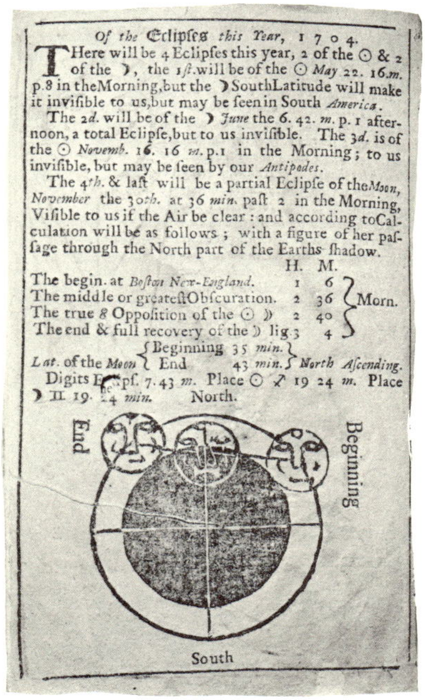

Of the Eclipſes *this Year,* 1704.

THere will be 4 Eclipſes this year, 2 of the ☉ & 2 of the ☽, the 1*ſt*. will be of the ☉ *May* 22. 16.*m.* p.8 in the Morning, but the ☽ South Latitude will make it inviſible to us, but may be ſeen in South *America*.

The 2*d*. will be of the ☽ *June* the 6. 42. *m*. p. 1 afternoon, a total Eclipſe, but to us inviſible. The 3*d*. is of the ☉ *Novemb*. 16. 16 *m*. p.1 in the Morning; to us inviſible, but may be ſeen by our *Antipodes*.

The 4*th*. & laſt will be a partial Eclipſe of the *Moon*, *November* the 30*th*. at 36 *min*. paſt 2 in the Morning, Viſible to us if the Air be clear: and according to Calculation will be as follows; with a figure of her paſſage through the North part of the Earths ſhadow.

	H.	M.	
The begin. at *Boſton New-England.*	1	6	Morn.
The middle or greateſt Obſcuration.	2	36	
The true ☍ Oppoſition of the ☉ ☽	2	40	
The end & full recovery of the ☽ lig.	3	4	

Lat. of the *Moon* { Beginning 35 *min.* / End 43 *min.* } *North Aſcending.*

Digits Eclipſ. 7.43 *m.* Place ☉ ♐ 19 24 *m.* Place ☽ ♊ 19. 24 *min.*

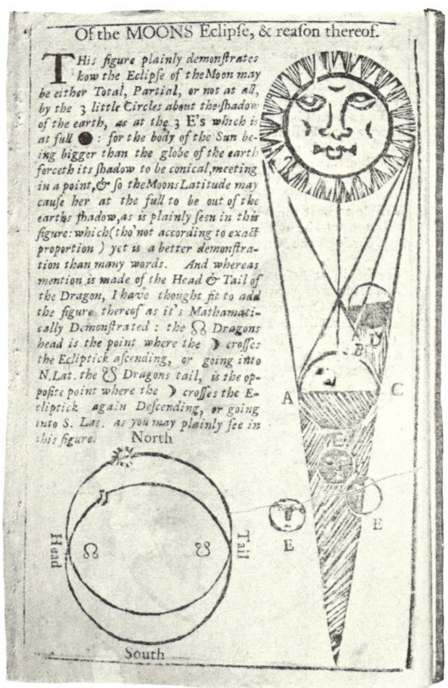

Of the MOONS Eclipſe, & reaſon thereof.

THis figure plainly demonſtrates how the Eclipſe of the Moon may be either Total, Partial, or not at all, by the 3 little Circles about the ſhadow of the earth, as at the 3 E's which is at full ● *: for the body of the Sun being bigger than the globe of the earth forceth its ſhadow to be conical, meeting in a point, & ſo the Moons Latitude may cauſe her at the full to be out of the earths ſhadow, as is plainly ſeen in this figure: which (tho' not according to exact proportion) yet is a better demonſtration than many words. And whereas mention is made of the Head & Tail of the Dragon, I have thought fit to add the figure thereof as it's Mathamatically Demonſtrated: the* ☊ *Dragons head is the point where the* ☽ *croſſes the Ecliptick aſcending, or going into N. Lat. the* ☋ *Dragons tail, is the oppoſite point where the* ☽ *croſſes the Ecliptick again Deſcending, or going into S. Lat. as you may plainly ſee in this figure.*

and schedules of courts, tables of money, and popular humor that too often was crude, coarse, and cruel.

As an especially useful source for local history, many almanacs contain unique registers of the state and city in which they were published, including not only the civic officers but also the schools, learned and moral and benevolent societies and institutions, mechanics' societies and fire companies, churches and hospitals, banks and insurance companies, and steamboat lines, attorneys and physicians. Such registers provide information on the organized life of communities that is difficult to assemble from other sources.

The almanac collection is about equal in its degree of completeness before and after 1820. Of the ninety-six American almanacs published before 1701, for which copies are known, the Society has fifty-nine originals and twenty-one bound photocopies.

The collection also includes most of the productions of the early eighteenth-century almanac makers—some of whom, in the popular mind of that day, were regarded as necromancers: Bowen, Clough, Holyoke, Whittemore, Robie, and Travis. Following these, through the later eighteenth and early nineteenth centuries, the New England states produced about twenty wonderful almanac series that are practically complete. Among them are scattered the more fleeting runs of four or six issues calculated by the ingenious amateurs of Connecticut and Massachusetts.

Other rich and distinctive groups of almanacs, each with its own 'personality' and each comprehensively represented at the Society, are the many elegant series printed at Philadelphia; the Pennsylvania German almanacs; those of early Pittsburgh and the Ohio Valley, keenly and promotionally regional; the often graceful and literary almanacs published from Delaware southward through Virginia and the Carolinas, with their gardener's calendars; and the illustrated almanacs of the 1870s, filled with lovely steel engravings.

The story of the formation of the almanac collection at AAS has been told in Brigham's *Fifty Years of Collecting Americana*. Some limitation has since been imposed upon its scope, and the collection now consists of 15,000 almanacs printed from 1656 through 1876 in the United States, Canada, Mexico, and the West Indies. About three-quarters of all the almanacs published in America through this period are at the Society, the only such comprehensive collection in existence. The Society continues to acquire lacking titles or issues with some regularity because almanacs from private collections continue to become available.

Full cataloguing of all of the Society's almanacs published through 1820 is completed. They are already entered in the Society's card catalogue under title and author or calculator, as well as by state and date, with full bibliographic descriptions. The catalogue entries for almanacs published through 1800, and for many of those through 1820, have been computerized and will eventually be available to research institutions throughout the country. Almanacs in the library printed from 1821 through 1849 are shown in Milton Drake's bibliography *Almanacs of the United States*, 2 vols. (New York, 1952), and those through 1876 are readily accessible by means of simple lists and files at the Society.

A special feature of almanac cataloguing at AAS is authorship attribution. Numerous almanacs were published anonymously or pseudonymously, but many of their calculators have been identified through close comparison with other almanacs for the same year and region. Identification of the authors of many of these has considerably expanded the work of known calculators.

The great size of the almanac collection has also provided a unique opportunity for assembling more knowledge about almanacs and their readers. As cataloguing has proceeded, two volumes of extracts, revealing what the matter in their calendar pages and other astronomical portions meant, what purposes it served, and how almanacs were made and published have been compiled. Eventually these will be arranged in permanent form to serve as a resource for the study of Americana unique to this library. [R.A.]

THE RESERVE COLLECTION

A rare book collection, such as that held by the American Antiquarian Society, serves not only as a source of research materials but also as a museum of a culture's printed documents. At AAS some of its treasures, but not all, are placed in the Reserve Collection. The purpose of the collection is, of course, to assure special handling of the books that have been assigned to it by the staff and by readers whose research *requires* the use of items of great rarity and, often, monetary value. Because the rarity, 'value,' or desirability of a particular item may lie in the eye of the beholder (being, in the case of the Society, its librarian) the selection of materials to be placed in Reserve is essentially subjective. Indeed, it must be because the extent and range of AAS collections is so far-reaching that it becomes impossible to segregate all worthy candidates for such an honor. In the past, only American imprints were placed in Reserve. However, with the discovery in our stacks of the London, 1677, edition in original vellum of William Hubbard's *Present State of New-England*, with the 'Wine Hills' map of New England intact, it seemed reasonable to afford special treatment to some American historical documents that have originated elsewhere than on this continent.

Monetary value may not be a reliable guide in making such determinations. Items that years ago sold for very little money have appreciated mightily in the market in recent years, not necessarily because they have become more useful as research documents, but, perhaps, because collectors have found them attractive as historical souvenirs. Nevertheless, this is a weighty consideration when assigning one or another rarity to Reserve.

The motivating factor, then, is importance. Thus, such books are admitted to Reserve as *The Whole Booke of Psalmes*,the Bay Psalm Book in its original vellum binding (Cambridge, Mass., 1640); *Pamela* by Samuel Richardson (Philadelphia, 1742), the unique copy of the first modern novel published in America; the first American edition of Milton's *Paradise Lost* (1777); *Specimen of Printing Types, from the Foundry of Binny & Ronaldson* (Philadelphia, 1812), the first example of American cast types; *History of the*

Expedition under the Command of Captains Lewis and Clark . . . to the Pacific Ocean (Philadelphia, 1814) in original, printed boards; *Tales for Fifteen* (New York, 1823) and *The Wept of Wish-Ton Wish* (Florence, 1829), the scarcest of the works of James Fenimore Cooper; *The Remarkable Story of Chicken Little* (Roxbury, Mass., 1840); Ovando J. Hollister's *History of the First Regiment of Colorado Volunteers* and *The March of the First, being a History of the . . . Regiment* (both Denver, 1863) among the earliest Colorado publications. These few titles are suggestive of the books that find their way into Reserve as the prime examples of their kind, ranging over the Society's entire collection. [M.A.McC.]

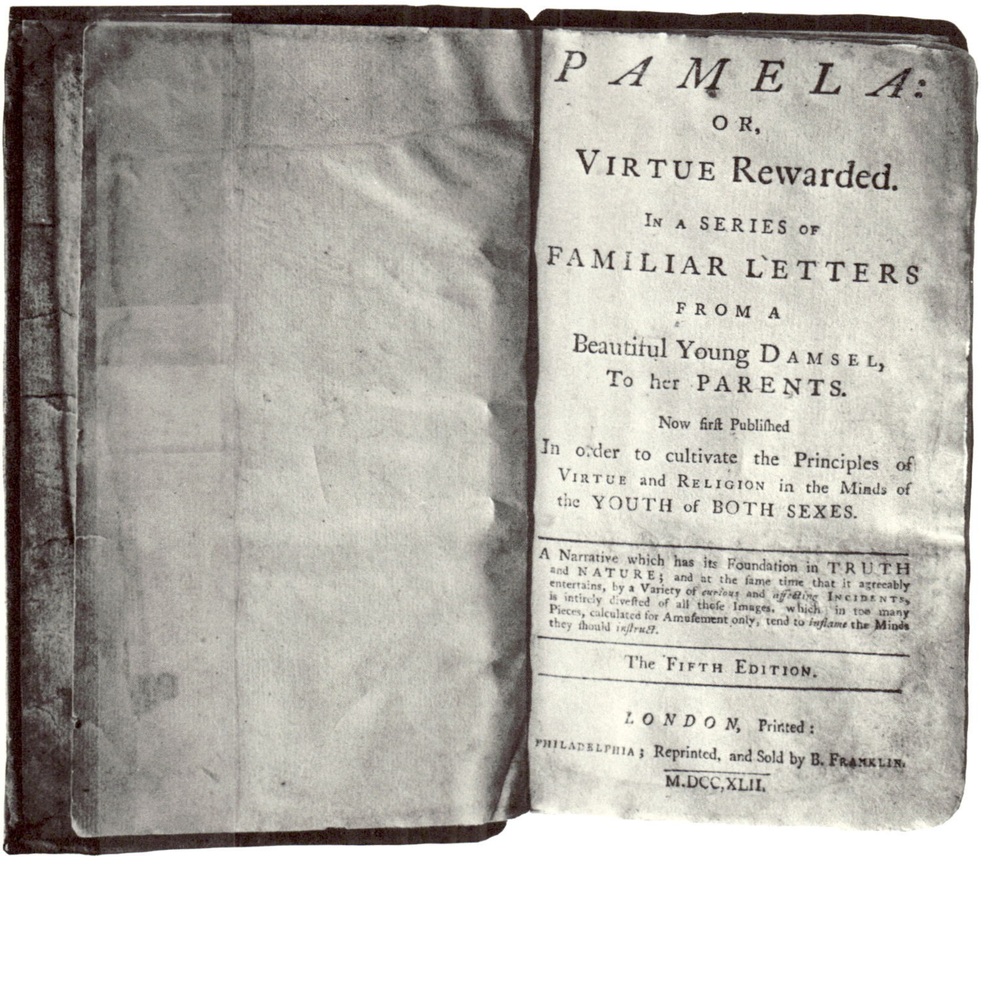

PAMELA:
OR,
VIRTUE Rewarded.
IN A SERIES OF
FAMILIAR LETTERS
FROM A
Beautiful Young DAMSEL,
To her PARENTS.
Now firſt Publiſhed
In order to cultivate the Principles of VIRTUE and RELIGION in the Minds of the YOUTH of BOTH SEXES.

A Narrative which has its Foundation in TRUTH and NATURE; and at the ſame time that it agreeably entertains, by a Variety of *curious* and *affecting* INCIDENTS, is intirely diveſted of all thoſe Images, which, in too many Pieces, calculated for Amuſement only, tend to *inflame* the Minds they ſhould *inſtruct*.

The FIFTH EDITION.

LONDON, Printed:
PHILADELPHIA; Reprinted, and Sold by B. FRANKLIN.
M.DCC,XLII.

BINDINGS

Another collection that is accorded special housing is that of books bound by American bookbinders. These volumes come from two major sources, Isaiah Thomas and Michael Papantonio. Thomas had the books in his library beautifully bound up for his own pleasure. Michael Papantonio, 150 years later, began to collect books bound in America during the seventeenth, eighteenth, and nineteenth centuries for his own pleasure. Not only are the books handsome to look at, but many are of outstanding importance as historical or literary works. The greatest binding in the house is that done by John Ratcliff for the Boston merchant Thomas Deane. It appears on Nathaniel Morton's *New-Englands Memoriall* (Cambridge, Mass., 1669) and came from the Papantonio Collection.

Guides to the bindings at the Society are found in *Early American Bookbindings from the Collection of Michael Papantonio*, 2d ed. (Worcester, 1985) and in Hannah D. French's *Bookbinding in Early America: Seven Essays on Masters and Methods* (Worcester, 1986). Additionally, the Society regularly publishes illustrated articles on bindings in its collections in issues of the *Proceedings*. The Bindings collection, when combined with the holdings of 'ordinary' bookbindings on the shelves in the general stacks, presents unparalleled opportunities to students of the American book arts. [M.A.McC.]

MINIATURE BOOKS

The earliest miniature books were produced primarily for personal convenience, to be carried in waistcoat pockets and ladies' reticules. As their popularity increased, the variety of texts expanded and, by the eighteenth century, included titles for children. The tiny volumes became exercises in printing and binding techniques.

Early American miniature books mirror their larger contemporaries: the texts are moral, the bindings plain. The majority of the volumes in the American Antiquarian Society's collection contain works of a religious nature. Of the 156 American editions of thumb Bibles cited by Ruth Elizabeth Adomeit in her *Three Centuries of Thumb Bibles: A Checklist* (New York and London, 1980), the Society holds ninety-eight examples, dating from 1765. The early examples of these abridgements of the Bible for children are in verse and, like the New England Primer, many of the same verses were included year after year. Similarly, the patriotic gesture was made in 1798 when Lower and Jones of Philadelphia dedicated their version 'to his excellency G. Washington, President of the United States of America.' A considerable number of the copies are worn and incomplete, mute testimony to hard use. Several of the volumes came with the bequest of d'Alté A. Welch, and Miss Adomeit has generously added to the collection.

All was not completely serious in the world of American miniature books. Songsters, hymnals, almanacs, and histories vied

LONDON
ROME
THE BIBLE 1819
Woodstock
BIBLE
BIBLE
BIBLE
BIBLE
Adomeit
THREE CENTURIES OF THUMB BIBLES
A Checklist
Garland
Mini.
Books
1852
[Soup]
pseud.
Life and
exploits
of Gen.
Scott.
New York
DICTIONARY
BIBLE

with entertaining and instructive texts. *Tom Thumb's Play-Book*, with Thomas's own inscription, 'Printed by I. Thomas when A'prentice in 1764 for A. Barclay,' is one of the Society's most valued miniatures. An educational title is *The Book of Nouns*, published in Philadelphia by J. Johnson in 1802. *London in Miniature: with 47 Engravings of Its Public Buildings* and *Antiquities and Costumes of Different Nations, in Miniature* were published in New York by Samuel Wood and Sons in 1816 and 1817. The familiar rhyme of *The Adventures of Mother Hubbard and Her Dog* was published in Albany in 1822 by G. J. Loomis & Co. and is found here in the original blue wraps.

Gen. Cass' Letter to the Harbor and River Convention (Chicago, 1848) and *Life and Services of Gen. Pierce* (Concord, Mass., 1852) are two gems of political satire. Miniatures produced by Frank Ellison of Waltham include *An Account of a Trip to the Sea Shore Made in the Year 1857*, and *A Journal of a Trip Down East, Aug. 1858*, a rare sporting miniature. Ellison kept a daily record of the weather for 1858–60 and printed miniature almanac pamphlets of meteorological tables and notes for the years. All of these may be seen at AAS.

A small group of twentieth-century miniatures is included in the collection. These consist of reprints of earlier titles and new titles, where the text is of historical interest. Among these are fourteen volumes produced by the late Achille J. St. Onge of Worcester.

There are approximately 350 miniature books in the Society's collection. They are shelved in chronological sequence by date of publication. Over the years, the size for inclusion in the collection has been arbitrary, with some volumes measuring over 3½ inches in height. Current policy is to limit the height of most new additions to 75mm, slightly less than 3 inches.

While there is no subject access to miniature books, there is a checklist of the entire collection, which is in two sections—an alphabetical listing by title and a chronological listing by date of publication. d'Alté A. Welch's *A Bibliography of American Children's*

Books Printed Prior to 1821 (Worcester 1972), contains some miniature editions and is annotated for AAS holdings, as is the Adomeit bibliography of thumb Bibles. [C.A.A.]

MISCELLANEOUS PAMPHLETS

The American Antiquarian Society holds approximately 38,000 miscellaneous, uncatalogued pamphlets published between 1831 and 1876. The importance of the pamphlet medium in nineteenth-century America may be overlooked by scholars, even though in fact, pamphlets are important because frequently they were generated as an instant response to governmental, political, educational, religious, and social affairs and events. The controversial issues and debates presented through pamphlets of that era reflect the extraordinary range, competence, and eccentricity of the authors, ranging from the very famous to the most obscure.

An estimated 15 percent of these publications are found to be unique and unrecorded in the *National Union Catalog: Pre-1956 Imprints*. Other items appear to be among only three or four surviving copies. The wide range of subject matter includes lectures, reports, essays, speeches, and treatises on education, social conditions, morals, politics, government, labor, industry, canals, railroads, agriculture, health and medicine, religion and theology, prison reform and management, temperance, and women's rights. The largest category appears to be sermons of all descriptions. Other significant genres include Fourth of July orations, congressional speeches, spiritualist messages, poetry, and collegiate lectures and inaugural addresses.

The collection also contains a large number of uncatalogued offprints of modern historical articles and other twentieth-century pamphlet material.

The collection is stored in acid-free boxes and arranged alphabetically by author, or chronologically for those lacking a personal author. Unfortunately, there is no subject access to these materials. However, those conducting research about a particular educator, politician, scientist, reformer, theologian, or other author may uncover a number of previously unknown publications by that individual. [C.A.A. and C.R.K.]

TOPICAL COLLECTIONS

THE BOOK ARTS

Bibliographies

The Bibliography collection includes about 1,000 volumes located in the Society's reading room. It consists of bibliographies of those topics, persons, regions, periods, and classes of publications in or about which the Society collects printed materials. It was assembled as a convenient reference collection for staff members, and in its public location it is equally available to readers, for whom it adds another dimension to the adjacent card catalogues. Also, many of the volumes in it are annotated by their compilers and often by the library's staff with the Society's symbol (MWA or AAS) beside those titles that are available at the Society.

Here are found all of the older general bibliographies of Americana, and the more recent state and local bibliographies: those of the old eastern and southern states, all of the midwestern states, California, and significant publishing centers such as Salem, Massachusetts, New London, Pittsburgh, and Lexington, Kentucky. Bibliographies for Hawaii and the West Indies are included, as well as bibliographies of Bermuda (in periodical literature) and of special regions such as the White Mountains and the Adirondacks. There is definitive coverage of the Great Plains and the Rockies, with attention to the Spanish Southwest.

Bibliographies of the major colonial and Victorian writers are here, along with others about such authors as Horatio Alger, Lafcadio Hearn, and Oliver Wendell Holmes. There are also bibliographies for nonliterary figures and topics such as Currier

& Ives, Paul Revere's engravings, and Benjamin Franklin's Philadelphia printing.

A few of the categories of publications for which there are bibliographies include atlases, city directories, nineteenth-century fiction, songbooks, penmanship and letter-writing books, children's literature, and American books of color plates.

A group of ancillary bibliographies from England are also part of the collection. These include bibliographies of British autobiographies, ballad entries in the Stationers' Register, catalogues of prints, drawings, and printed music in the British Museum, and the seventeenth-century descriptive catalogues by London booksellers.

Among the classes of works in the Bibliography collection with greatest potential interest for readers are the topical listings, which cover such subjects as the circus and allied arts, American fishing books, early American sports, medical imprints, popular music, anti-Masonic literature, travel books, and technical publications prior to 1831. Also located here are the Library of Congress catalogue of naval records of the American Revolution, Joseph Smith's great descriptive bibliography of writings by members of the Society of Friends, a twenty-five-volume Baptist bibliography, and three shelves of bibliographies about the newspapers of states and cities.

Finally, adjacent to the Bibliography collection are the national catalogues. These are, respectively, the *National Union Catalog* and the British Museum's *General Catalogue of Printed Books.* The *National Union Catalog* was first published as the *Library of Congress Catalog,* reproducing in book form the author and other main-entry printed cards in the LC card catalogue. Now, as the NUC, it incorporates with these the cards contributed by most of the larger United States libraries through 1955, with a twenty-year supplement. The British Museum catalogue reproduces, in 313 volumes of abbreviated entries, the book-form catalogue of that institution. [R.A.]

Printing and Publishing History

The books at either end of the large Z classification at AAS, denoted Printing and Related Arts, suggest how relevant its section is to some of the key interests of the Society. The first book on the shelf in the Z section is William Loring Andrews's *Bibliopegy in the United States and Kindred Subjects*, a lovely book published in 1912. Its presence there reminds us of the Society's preeminence as a center for the study of the bookbinder's craft in America, a position solidified by the bequest to AAS of the Papantonio Collection and the Society's recent publication of two books of important bibliopegical research. The last book within the section is a salesman's sample copy of the *Young People's New Pictorial Library*. It brings up the rear of a large collection of book salesmans's samples, or blads, some of it recently acquired. This interesting volume, together with the others in that sub-collection, reminds us that the study of the commerce in books, no less than

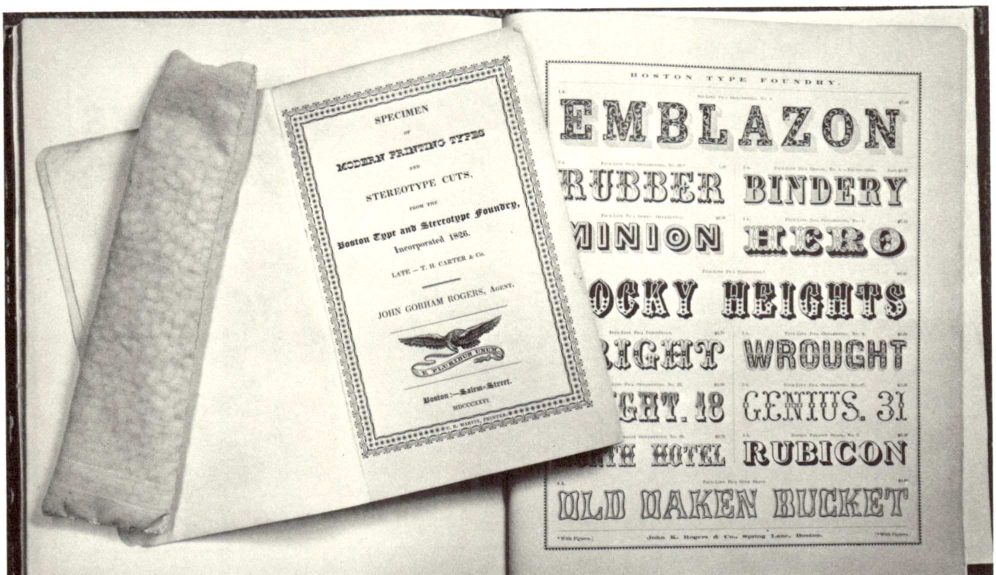

of the printing of them, is an important aspect of the work of the Society's Program in the History of the Book in American Culture.

The books in the Z section are primarily secondary materials, covering virtually all aspects of printing, publishing, and related activities, although they also include materials that may be considered primary—even some manuscript material and some items, like the salesman's dummies, that have special artifactual value. A listing of the principal subdivisions within the classification indicates its range: bookbinding, American journalism, foreign journalism, magazines and periodicals, paper, printing techniques and practices, publishing and bookselling, and writing. The largest groups are books on the histories of printing, publishing, and bookselling. There is good coverage of the British side of the subject, which is useful to place the history of the book in America in its cosmopolitan context.

Some of the more interesting items in the collection are those that may be considered primary materials. They include a small collection of nineteenth- and twentieth- century press association reports, manuals on how to be a printer, publisher, or journalist, a sizable collection of type-specimen books, the aforementioned salesman's dummies, and six boxes of receipted bills, arranged alphabetically by state. These latter items are receipted bills from various American newspapers for payment for subscriptions or advertising. Some are wholly manuscript, while others are filled-in printed forms. A few date from the late eighteenth century, but most from the nineteenth. The great majority are from Massachusetts, particularly Boston and Worcester.

The materials in the Z section are fully catalogued in the General Catalogue. [J.B.H.]

Book Salesman's Samples

The Society holds a particularly strong collection of salesman's samples or dummy books, which have also been called blads. These specimen books were used by the itinerant agents of sub-

scription publishers to entice customers to subscribe to an edition, that is to agree to purchase a copy of a work in advance of delivery. This form of marketing was common in America during the years just before and after the Civil War. Unlike the phenomenon of the same name in the eighteenth century, nineteenth-century subscription publishing was used to finance or at least to assure the success of an edition prior to printing.

A typical specimen book of the nineteenth century consists of sample sheets in a sample binding. Sometimes these sheets are in a prepublication state and thus may have bibliographic interest. A customer could choose among alternative styles of binding, and a sample book usually includes the spines of available styles mounted inside the covers. The sample book might also include a printed prospectus describing the virtues of the work or edition with recommendations from well-known persons. Most specimen books also include blank forms on which the salesman entered the names, addresses, sometimes their occupations, and the number of copies his subscribers agreed to purchase. Subscription forms can provide demographic information on book ownership, if not on actual readership.

The Society holds approximately 350 salesman's sample books issued before 1900; about one-third of these were issued before 1877. The earliest was issued in 1836. Most of the early ones date from the 1860s and 1870s. The collection includes dummies of four first editions of works by Mark Twain. Sample books issued before 1877 may be found in the General Catalogue under 'Prospectuses.' A checklist is available for those issued after 1876.

Complementing this collection are collateral materials on the subscription business, including several agents' manuals that give explicit instructions on sales methods and biographies and autobiographies of book agents. These materials can be found listed in the General Catalogue under 'Booksellers and bookselling.'

[R.C.F]

Booksellers' and Auction Catalogues

Another collection important to the history of the American book trades is that of booksellers' and book auctioneers' catalogues. They are shelved on three ranges of book stacks and include examples from the beginnings of the American wholesale and retail book trades. Catalogues issued prior to 1821 are fully catalogued in the Society's Imprints Catalogue. Those that appeared in 1821 and to the present are filed alphabetically on the shelf by firm name. They are not catalogued.

The Society published Robert B. Winans's *Descriptive Checklist of Book Catalogues Separately Printed in America, 1694–1800* in 1981. This work and George L. McKay's *American Book Auction Catalogues, 1713–1934* (New York, 1937), introduced by Clarence S. Brigham, provide annotated guides to AAS holdings.

[M.A.McC.]

Bookplates and Booksellers' Labels

The bookplate collection is arranged in two segments. One is arranged alphabetically by the last name of the owner; the other collection, institutional plates, is arranged alphabetically by the name of the city in which the institution is located. The collection was started by the Reverend Herbert E. Lombard in 1915, and very substantial acquisitions occurred within the following decade. Additions to the collection are made quite frequently. When a book with a new bookplate is acquired, a photocopy is made and inserted into the collection with the provenance of the bookplate. When the Society acquires duplicate volumes, any bookplates are lifted from the duplicate volumes and added to the collection. There is no index to the collection by designer, engraver, or printer, although Charles Dexter Allen's *American Bookplates* (New York and London, 1894) does have a listing of eighteenth-century bookplates that are signed by the engraver. There is an annotated copy of this study housed with the collection of bookplate literature.

A recent addition to the bookplate collection was the gift of Dorothy Sturgis Harding, consisting of preliminary drawings, proofs, miscellaneous sketches, and some printing blocks for the plates that she herself had designed. Her collection joins similar ones of Edwin D. French and Sidney Smith and others that were described by Clarence Brigham. Although Brigham suggested that AAS would continue to collect twentieth-century plates, this has not been the case.

Another collection that grows in the same manner as the bookplate collection is the group of booksellers' labels. Arranged alphabetically on index cards, these labels (over 1,400 of them) have been acquired from collectors and book dealers over the years. In recent years, it has become routine to photocopy any labels or binders' tickets in volumes and add these to the file, along with information on the provenance of the label. This small but expanding collection provides a source for the names of binders in major urban and rural centers and serves as an index, albeit a very partial one, to signed bindings in the Society's collection. It also includes references to trade cards. A number of these labels were reproduced in volume 82 of the *Proceedings* (1972) with a brief introductory note. [G.B.B.]

THE UNITED STATES

American History and Culture

In the scheme devised by Clifford K. Shipton to organize the Society's collections, the E and G classifications contain the core of the Society's post-1820 holdings. The E classification provides a chronological arrangement of works on United States history, the G classification a topical arrangement. Each classification contains both primary and secondary materials. Recent estimates indicate that the E class contains 12,500 items, the G class 28,000.

In the E classification, which Shipton conceived as a simplified version of the Library of Congress's E classification, sections are arranged under both general and limited headings, providing what is, in effect, a chronological outline of United States history. Broad topic headings (e.g., E150, the Era of the Revolution, 1764–83) are followed by narrower topic headings (e.g., E170, the Boston Massacre; E220, the Declaration of Independence; E275, Loyalists). In addition to the chronological portion of the classification (E100–E675), space is provided for works whose focus is regional (e.g., E45, New England; E50, the Ohio Valley). The largest of these, E85, is devoted to the West (Plains and Rockies) and contains much of the collection of western Americana given to the Society by Donald McKay Frost.

As noted above, the E classification contains both primary and secondary materials. Since American printed materials for the period through 1820 are located in the imprints collections, this part of the classification generally contains secondary materials, together with some early British and European imprints that pertain to the subject. However for the period 1821–76, E is a

ASHER BENJAMIN
BENJAMIN'S ARCHITECT
BENJAMIN
THE COUNTRY BUILDER'S ASSISTANT
BENJAMIN'S ARCHITECT
DESIGN Nº 6.

trove of historical materials, well supported by secondary works. The collection of primary materials from the Civil War period is especially strong, and includes an excellent collection of regimental histories. The Society is necessarily selective in acquiring modern historical writing on the Civil War, purchasing only the best of the hundreds of titles published on the subject each year. The Society's holdings relating to slavery are also particularly strong, numbering over 2,000 titles.

The G classification accommodates works addressing particular topics in American history. In this classification, Shipton prepared an alphabetical list of 140 topics and assigned each topic a number in the G class. His list reflects AAS holdings and is not an attempt at a hierarchical or universal scheme. Topics range from Advertising (G100), Aeronautics (G120), and Agricultural History (G130), through such topics as Indian Languages (G465), Phrenology (G673), and Transportation (G840), to Whaling (G900), the Whig Party (G925), and Zoology (G975). As with the E classification, both primary and secondary works are included. Works of Fiction (G526) and Poetry (G850) are included in this class; (G526) is the largest section in G.

Access to materials in the E and G classifications is provided in the Society's General Catalogue. The quality of this cataloguing varies considerably. While every item is at least nominally catalogued, some items are represented by no more than the briefest of notes at the foot of a card describing an earlier edition. Many items are represented by a single card, usually (but not always) under author. Subject access often extends no further than a single entry under a broad heading. Other items are fully catalogued. Unfortunately, modern secondary works are likely to be well catalogued, while uncommon primary materials may be represented by cataloguing that is nearly as old as the materials themselves. In an effort to remedy this deficiency, the Society is gradually undertaking the retrospective conversion of its older cataloguing to machine-readable form. Thus far, the Society has recatalogued in RLIN its works on Architecture (G150), its collections on Slavery and Afro-Americana (E455, G600), and the preeminent collection of editions of the works of James Fenimore Cooper (G526). [A.N.D.]

Government Documents

At the third annual meeting of the American Antiquarian Society on October 24, 1814, Isaiah Thomas advised members that 'every measure that can be adopted to make the Society appear respectable as a national institution must be desirable.' To achieve such stature, he suggested that members petition the federal government 'to send the laws of the national government to be deposited and preserved in our library.' Such action was taken, and this is reflected in the daily accounts of the debates and proceedings of the thirteenth Congress. On December 1, 1814, Congress approved 'Resolution 7: For furnishing the American Antiquarian Society with a copy of the journals of Congress; and of the documents published under their order.' This special act of Congress had the effect of making the Society the first depository library in the United States for federal documents (other than the Library of Congress).

Isaiah Thomas requested the legislatures of the several states to make provisions for deposit of their publications at AAS. Records from cities and towns were also solicited. Thomas's ambition has been fulfilled; today the Society holds one of the nation's finest collections of early American government documents.

The American State Papers and the nineteenth-century U.S. Serial Set form the basic collection of federal documents. The American State Papers contain reprints of the documents of the first fourteen Congresses (1789–1817), arranged by subject class in thirty-eight volumes; these are generally considered to be a part of the Serial Set. The Sheep-Bound (so called because of its early bindings) or Serial Set consists of congressional journals, reports, and related internal publications; executive branch material, including presidential messages and administrative reports of departments and agencies; and miscellaneous documents from independent bodies or commissions that were printed by order of Congress. The Society's set is virtually complete through 1876 and reasonably complete to the end of the nineteenth century, and in some instances, the set extends into the twentieth century. There is a massive body of congressional material, as well as annual reports and series publications from the Department of the Interior and its many bureaus, the Smithsonian Institution, and the Department of State, to name but a few. Miscellaneous sets issued by acts of Congress for independent bodies include *American Archives* (Washington, D.C., 1837–51), a documentary history of the American Revolution edited by Peter Force; *The War of the Rebellion: A Compilation of the Official Records of the Union and Confederate Armies* (Washington, D.C., 1880–1902), and M. C. Perry's richly illustrated *Narrative of the Expedition of an American Squadron to the China Seas and Japan, Performed in the Years 1852, 1853, and 1854* (Washington, D.C., 1856).

As in most libraries, the Serial Set is arranged according to the Superintendent of Documents (SuDoc) classification system. This classification system is used as a method of identification in SuDoc bibliographies and lists, as well as lists issued by various government departments and agencies. Several partial indexes provide access to documents in the set. The most important of these for the Society's holdings is the *Checklist of United States Public Documents 1789–1909* (Washington, D.C. 1911), which is fully annotated. Other major indexes to the nineteenth-century Serial Set include B. P. Poore's *A Descriptive Catalogue of the Government Publications of the United States, September 5, 1774–March 4, 1881* (Washington, D.C. 1885), A. W. Greely's *Public Documents of the First Fourteen Congresses 1789–1817* (Washington, D.C. 1900), and J. G. Ames's *Comprehensive Index to the Publications of the United States Government 1881–1893* (Washington, D.C. 1905). The 1977 publication of the *CIS U.S. Serial Set Index* provides excellent subject access to material in the set. The Society has part 1 (1789–1857) and part 2 (1857–1879) of the important finding aid, published by the Congressional Information Service, Inc.

AAS also holds an impressive collection of separately issued early federal documents. These include congressional journals,

Smithsonian Year · 1977
National Endowment for the Humanities
Nineteenth Annual Report
1984
LAWS
1727
VIRGINIA
ANNUAL REPORT OF THE LIBRARIAN OF CONGRESS · 1969
1606

committee reports, presidential messages, treaties, laws, proclamations, and other official decrees. Those issued through 1820 are fully catalogued and are accessed through the Imprints Catalogue under corporate main entry (United States), and added entries and subject headings. These are integrated in the Dated Books and Dated Pamphlets collections. Federal documents issued after 1820 have been catalogued selectively. Documents in the Broadsides collection printed through 1830 are fully catalogued and may be accessed through RLIN. Additionally, federal publications are found in the Frost Collection, the Reserve Collection, and U.S. Records Collection, 1777–1876, in the Manuscripts Department.

As a member of the depository library program, the Society still selects some current documents from the *List of Classes of United States Government Publications Available for Selection by Depository Libraries*. While the number of classes selected for deposit is small (172 from the 6,509 depository item numbers available were selected as of June 1986), the categories represent the collecting interests of the Society in areas of history, bibliography, and subjects in which it is already strong. Although the selected classes cover a variety of government bodies, the primary sources for these documents are the Department of the Interior, the Library of Congress, and the Smithsonian Institution. Since 1977, a shelf-list of current depository items has been maintained according to Superintendent of Documents classification numbers. Occasionally, current publications are catalogued and integrated with the appropriate subject collection.

The Society has a representative selection of state and municipal documents issued through 1876 from all regions of the nation, and these provide an excellent body of material for historical and early legal research. Publications from the New England states, including legislative journals, session laws and statutes, are nearly complete through 1820 in their original editions. The Society holds many of the early laws and journals of New York State, as well as regions of the South and West, such as early acts and legislative journals of Kentucky, Tennessee, Illinois, and Indiana.

State publications also include legislative manuals, annual reports on agriculture, commerce, and transportation, and various series publications such as that on the natural history of states.

The collection of municipal documents is especially strong for New England, although partial holdings do exist from all regions through 1876. There are municipal reports ranging from San Francisco to Portland, Maine, and from St. Louis to New Orleans. The Society has a fine collection of city and town ordinances and codes for New England and the central states; original or early editions of the charters of major cities; and a wide variety of annual reports relating to local matters such as education, public health, property assessments, crime, and recreation. Many municipal documents contain lovely illustrations. For example, New York City's *Annual Report of the Board of Commissioners of the Central Park* has important information on the design, description, and expenditures for the park, as well as superb plates of maps, engravings, lithographs, and early photographs of Central Park and its environs. The Society has a complete set of these fourteen reports, which were issued from 1857 to 1871.

State and municipal documents issued through 1800 are fully catalogued; those printed between 1801 and 1830 are partially catalogued. The Imprints Catalogue and RLIN provide access to the collection under appropriate jurisdiction (e.g., Connecticut; Concord, Mass.; etc.), as well as access by added entries and subject headings. These documents are integrated on the shelves throughout the library in the same manner as the federal documents. State and municipal publications issued after 1830 are uncatalogued, except on a very selective basis.

There are no specific bibliographies to ease access to state and municipal documents. State publications held by the Society are partially annotated in *The Catalogue of the Library of the Law School of Harvard University* (Cambridge, Mass., 1909). The uncatalogued sets of state and municipal documents generally have locally produced indexes to a particular series that are helpful in locating specific items. For example, the volume of *Boston City Documents for 1875* includes a useful general subject index for their publications issued from 1834 to 1874. When attempting to locate any government documents, whether catalogued or uncatalogued on the national, state, or local levels, users should be certain to consult with the staff of the readers' services department.

The Society continues to add to its preeminent collection of early American documents. Although separately issued works are scarce in today's market, AAS has recently acquired important government publications. For example, in 1974 a generous gift from the former First Federal Savings and Loan Association of Worcester, Massachusetts, enabled the Society to purchase 170 separately printed laws issued by the first four Congresses of the United States. This rare collection of 'slip laws' included drafts of bills, texts of treaties, and proposed constitutional amendments. Known as the First Federal Collection, this uncommon group of working papers includes the first American copyright law, the first printing of the 'Bill of Rights,' the first federal law regulating trade with the Indians, the first labor legislation, and the first law establishing a uniform rule for naturalization of citizens.

In effect, the Society serves as a depository for eighteenth- and nineteenth-century government documents in a twentieth-century environment, much as Isaiah Thomas envisioned in 1814.

[J.D.C.]

Local, County, and State Histories

Local histories are a literary manifestation of an American self-consciousness and have been written and read since William Bradford set down his account of the establishment of Plymouth Plantation. Significant contributions to the genre were made after the centennial and again in the 1930s, when the Federal Writers Project produced the first uniform series of state histories. The field has recently been enlarged by the many state and local histories that were written to commemorate the American Revolution bicentennial. Traditionally, the field has been dominated by amateur historians, but since World War II professional historians have contributed increasingly to both the monographic and periodical literature that constitutes local history.

The Local History collection at the American Antiquarian Society encompasses more than 55,000 volumes published between 1831 and the present. Local histories published before 1831 are part of the Society's Dated Books and Dated Pamphlets collections, and many rare examples of local history are shelved in the Reserve Collection. It is among the largest and most frequently used resources at the Society, and is the one collection that best reflects the Society's national scope, for it includes works of all fifty states and for thousands of their constituent counties and local communities. The collection is shelved alphabetically by state, with statewide histories followed by alphabetic arrangements of county and then city and town histories. In addition to the straightforward histories, the Local History collection includes narratives of description and travel, gazetteers and dictionaries of place names, bibliographies of local imprints, institutional histories, biographical encyclopedias, and genealogical records that include vital and census statistics, abstracts of wills, and transcriptions of cemetery inscriptions. Also in the Local History collection are the publications of such state and local historical agencies and societies as the Essex Institute, the Pennsylvania German Society, the South Dakota Department of History, and the Society of California Pioneers. All of these materials are accessible through the General Catalogue, with at least one subject entry provided under the state, county, or local place name. More recent cataloguing efforts also provide access by author, editor, and title.

Some uncatalogued materials are also part of the collection. For example, the Society's 'Worcester Collection' includes a large number of uncatalogued pamphlets and clippings relating to the city. (These materials were organized in 1979–80 with an accompanying checklist.) Approximately 250 periodicals such as the *Virginia Magazine of History and Biography*, the *Chronicles of Oklahoma*, the *Dukes County Intelligencer*, and the *Atlanta Historical Journal* are also shelved with the Local History collection. Access to these periodicals is provided through an index maintained by the serials department.

Because of the number and variety of works that fall under the heading of local history, there is no standard reference work for the field. However, since the category of regional, state, and local history is one of the six sections included in *America: History and Life* (Santa Barbara, 1964–), this index, located on open shelves in the reading room, is an invaluable source for local history periodical literature. Another useful reference source is the six-volume *Bibliographies of New England* series. Prepared by the Committee for a New England Bibliography, and published between 1976 and 1986, this series represents a recent and successful effort to extend bibliographical control over the vast historical resources of this region printed from the earliest time to the present. The bibliographies contain entries for both monographic and periodical literature in the areas of political, economic, social, and intellectual history. Entries are alphabetical within a geographical framework; works relating to the state as a whole or to several counties appear first, followed by works relating to single counties, and then by those relating to cities and towns. A third reference tool is *A Bibliography of American County Histories* (Baltimore, 1985), compiled by P. William Filby. It lists 5,000 county histories published to date and includes information about reprint editions and separately published indexes. This information is of particular importance to AAS because the acquisition of indexes makes any of the originally unindexed histories even more valuable to scholars, genealogists, and staff alike. And the addition of reprint editions to the collection allows the originals to be withdrawn from circulation and preserved as examples of American printing history. [D.N.O.]

Western Americana

The American Antiquarian Society holds an extensive collection of materials dealing with the Trans-Mississippi West: histories of discovery and exploration, guides for settlers, descriptions of the countryside, and expositions of the manners and modes of living in the frontier West. The core of these holdings is the Donald McKay Frost Collection of more than four thousand titles that Frost acquired over thirty years. His collection, given to the Society in 1947, includes not only overland narratives but also town and local histories and biographies of early settlers. Frost was an ardent collector and tried to purchase all the rare things offered him, including all editions of a work. AAS lacked 1,775 of the titles Frost had acquired but already owned some of the early

titles and also a large number of government imprints and literary titles that had not been of interest to Frost.

Many of the Frost books are in excellent condition. The collection contains some association copies bearing author's signatures. Among items of interest are memoirs of the Lewis and Clark expedition; personal recollections of Zebulon Pike; navigators for rivers, including descriptions of towns, ports, and harbors; accounts of the Indian land titles and treaties; and emigration guides for journeying through, and settling in, the American West. Among the more notable items in the collection are *History of the Expedition under the Command of Captains Lewis and Clark* (Philadelphia, 1814) and John Ledyard's *A Journal of Captain Cook's Last Voyage to the Pacific Ocean* (Hartford, 1783).

The Society holds about 35 percent of the most important titles listed in the Henry R. Wagner-Charles L. Camp bibliography, *The Plains and the Rockies: A Bibliography of Original Narratives of Travel and Adventure, 1800–1865* (San Francisco, 1982).

Items dealing with the development of the West may be found in other collections at AAS as well. Researchers will find relevant materials in Dated Books and Dated Pamphlets; Local, County, and State histories; Newspapers and Serials; Graphic Arts (for maps); and in the Biography collection. Access to all this material is through the card catalogues. In addition, the acquisitions staff annotates the 1937 edition of Wagner-Camp's *The Plains and the Rockies* to reflect additions to the Society's collection of Western Americana.

The Society continues to add to this collection through purchases made on the Donald McKay Frost Fund, which was created by the bequest of that outstanding collector. [S.J.W.G.]

The Hawaiian Collection

The collection of printed materials from and about the Hawaiian Islands is representative of the print culture of these mid-Pacific islands as impressed upon the native population by the American missionary movement of the nineteenth century. The collection includes bibliographies, biographies, Hawaiian history, Hawaiian language imprints, maps, newspapers, periodicals, and engravings. The extent and variety of the collection permit research about the eighteenth- and nineteenth-century political and cultural history of the Sandwich Islands as they became the Hawaiian Islands, the impact of the American missionary movement on an indigenous culture, as well as the study of the dissemination of American printing technology outside the continental borders of the United States.

The collection of Hawaiian materials at AAS began with the materials deposited in the mid-nineteenth century by one of the American Congregational missionaries, the Reverend Mr. Samuel C. Damon, who also was a member of the Society. However, the major thrust of the collection derived from the purchase, in 1937, of the Hiram Bingham Library. This acquisition was made possible through the interest and generosity of two descendants of American missionaries, Foster Stearns and James M. Hunnewell. While comprehensive in its coverage and broad in its scope, the Hawaiian collection at the Society is not as large as the collections housed on the islands themselves at the Hawaiian Mission Children's Society Library or at the Bishop Museum. In the United States, the major collections, in addition to that at AAS, are held at the Houghton Library at Harvard and the Newberry Library. The Society's Hawaiian collection includes 190 volumes of Hawaiian imprints from about 1820 to 1896. A major portion of these materials were printed on the hand presses sent to the islands by the American Board of Commissioners for Foreign Missions in the Pioneer Company in 1820 and in the Third Company in 1828. The incunabula for Hawaiian printing could be considered those materials produced in the period 1822–29 chiefly on the island of Oahu; the earliest printed piece is *The Alphabet*, printed in Oahu on the Mission Press in January 1822. Other printing locations represented in the collection are from the islands of Maui and Hawaii, as well as New York.

Closely supporting the imprint collection is the extensive accumulation of more than 580 bound volumes, pamphlets, and the reports of the Hawaiian Historical Society, which provide historical perspective on the islands and the missionary movement. Most of this collection of material is filed together in the Society's Local History section.

Of equal importance to the collection is the broad array of newspapers and periodicals originating in the Hawaiian Islands. There are twenty-two English and twelve Hawaiian-language titles in this collection. Most of the newspapers are listed in Gregory's *American Newspapers, 1821–1939* and are now being included in the CONSER newspaper file, mounted on the OCLC system, as well as in the RLIN file.

One of the more unusual portions of the collection is the assortment of more than thirty engravings produced by students at the Lahainaluna School on the island of Maui. A mission press was introduced at this institution about 1828 and was used to provide male students with instructions in the skills of engraving and printing. No complete inventory of Lahainaluna engravings has been made, but the number reported in various locations exceeds 100. A checklist made by George T. Lecker in 1927 records thirty-three maps and fifty-seven sketches of houses and landscapes, only one of which is of a non-Hawaiian subject. Of interest to residents of the greater Worcester, Massachusetts, community is the fact that it is a view of the town common of Holden, Massachusetts, circa 1840, as sketched from memory by Edward Bailey, a teacher at the school and a native of Holden.

Bibliographic control of Hawaiian-language material was for years dependent upon an unpublished inventory compiled by Howard M. Ballou in 1908. However, through the urging and support of the late Clarence S. Brigham, a new and greatly expanded bibliography of Hawaiian imprints was begun in 1938 by Bernice Judd, librarian at the Hawaiian Mission Children's Society. In 1963, Janet E. Bell updated and revised Judd's work, and in 1971, with the assistance of Clare G. Murdock, she was able to complete a new volume, *Hawaiian Language Imprints, 1822–1899: A Bibliography*. The volume was published by the Hawaiian Mission Children's Society and the University Press of Hawaii.

The collection of Hawaiiana at the Society includes 136 of the titles listed in this bibliography. [F.E.B.]

Biography

The American Antiquarian Society's collection of almost 30,000 biographies is a major resource for scholars. Generally speaking, the biographies represented in the collection are of individuals who were resident in the United States or Canada and who were born before the year 1851. Although the cataloguing of this material (in the General Catalogue) is similar to that in other research libraries (by author and by subject of the biography), the collection itself is distinctive in the rich variety of its holdings, since it includes sources generally considered ephemeral: funeral orations, commemorations, masonic tributes, and privately printed tributes of individuals not sufficiently well known to warrant more standard biographies. There are also collective biographies in the Society's holdings, including standard sources such as the *Dictionary of American Biography* as well as more unusual sources like *Memoirs of Several Deceased Members of the New-England Historic, Genealogical Society* (Boston, 1878) and *The Life and Sketches of Curious and Odd Characters* (Boston, 1840). A few collective biographies, such as William B. Sprague's multivolume *Annals of the American Pulpit* (New York, 1857–69), have been indexed by Society staff to provide easy access to individual entries.

[C.A.A. and N.H.B.]

Genealogy

The American Antiquarian Society has a very strong collection of published genealogical material focusing on early North American lines of descent, including French-Canadian genealogies. Not only is this collection used by genealogists but it is also used extensively by scholars working on biographical, historical, and literary topics. As of 1987, the collection numbers over 17,000 family histories, plus 2,000 genealogical reference works. The Society adds to the collection by purchase and donations.

The collection includes some of the earliest genealogies published in the United States: Luke Stebbins's *Genealogy of the Family of Samuel Stebbins* (Hartford, 1771), John Farmer's *A Family Register of the Descendants of Edward Farmer* (Concord, N.H., 1813), Anthony Haswell's *Record of the Family of Anthony Haswell* (Bennington, Vt., 1815), and Joseph Sharpless's *Family Record of the Sharpless Family* (Philadelphia, 1816). More information pertaining to the Society's collection of early genealogies can be found in volume 32 of the *Proceedings of the American Antiquarian Society* (1922).

All the family histories in the Society's possession are fully catalogued (in the Imprints Catalogue), with access by author and by all the major families traced in the book. The genealogical reference materials also are catalogued, with many of them available on open shelves in the reading room. The reading room collection includes the 240-volume series of vital records of Massachusetts towns to 1850, the *New England Historic Genealogical Register*, lineage books published by the Daughters of the American Revolution, *Massachusetts Soldiers and Sailors in the Revolutionary War*, Mayflower materials, and P. William Filby's *Passenger and Immigration Lists Index* (Detroit, 1981). Of particular help is the *American Genealogical-Biographical Index* (Middletown, Conn., 1952–). This series indexes 631 family histories by personal names. The Society owns 591 of the titles indexed, and, to assist reader usage, the consolidated 'Key Title Index' to this series has been annotated with the library's call numbers. One reference that is of particular interest for genealogical research in Worcester County is the Works Progress Administration's 'A Biographical Index to Worcester and Worcester County' (typescript, 1935). This work indexes ten standard histories by the personal names of the major figures who lived in this area.

Bibles with manuscript notes entered by family members are an excellent and often overlooked source of information for

GENEOLOGICAL FAMILY TREE OF JOHN GEORGE BECHTEL.

genealogists. The Society has a collection of these Bibles, with access provided in the catalogue by family name.

The Society has a wide spectrum of auxiliary materials available for genealogical research. These include state, county, and local histories for all fifty states, biographies, United States histories, regimental histories, Canadian histories, periodicals of local historical societies and state libraries, and city directories. Another excellent source of genealogical information is the Society's collection of newspapers. The Society's typescript 'Index to Marriages in *Massachusetts Centinel* and *Columbian Centinel*, 1784–1840,' and the 'Index to Deaths in *Massachusetts Centinel* and *Columbian Centinel*, 1784–1840,' are two widely used indices that are available in the reading room.

The Society also owns a set of clippings of the genealogical column 'Notes and Queries,' which ran in the *Boston Transcript* from 1894 to 1941. This is a very good source for genealogical information pertaining to New England for the period 1600–1800. The Society's collection of these articles is arranged in boxes by date of publication, and indexed by name in the *American Genealogical-Biographical Index*.

There is a small collection of genealogical materials available in microform. In addition to the microform series *Early American Newspapers*, by Readex Microprint Corporation, there are two useful microfiche series that are of importance to genealogists: Jay Mack Holbrook's *Massachusetts Vital Records*, which continues the earlier published records up to 1890, and *American Directories*, produced by Research Publications, Inc., which includes all the directories listed in Dorothea N. Spear's *Bibliography of American Directories through 1860*, published by AAS in 1961.

In order to derive maximum benefit from the Society's genealogical collections, readers are encouraged to do preliminary work before using the collections. [E.S.A. and M.E.L.]

Directories of Localities

As the United States developed, people began to search for means whereby they could obtain information about their fellow citizens and their community. The first American directory was published in *The South Carolina and Georgia Almanac for . . . 1782*, a listing of 304 names indicating the trades or professions and addresses of residents of Charleston, South Carolina.

The first separately published directory in the United States was *Macpherson's Directory for the City and Suburbs of Philadelphia*, published in October 1785, which was followed a month later by the publication of Francis White's *The Philadelphia Directory*.

Early directories were modest forerunners of later publications. Names were listed alphabetically by the first letter of last names but, under a given letter, the sequence was in order of the importance of the citizen. Thus, in *The Boston Directory* of 1789 the first name listed was 'Samuel Adams, Hon. Winter St.', while at the end of the A's appeared 'Samuel Adams, Truckman, Eliot St.' Canvassers for these early publications had to list locations before there were street names or numbers. In *The Directory for the City of Hartford for . . . 1799*, 'David Brewer & Co., druggist,' was located 'at the dwelling house of John Balles,' and 'John Balles, Tavern keeper,' was found at the 'Sign of the "Lamb," North end of city.'

Various kinds of information were included in the early directories. For example, *The New-York Directory, and Register* and *The Boston Directory* of 1789 both included an early map or plan of their respective cities. In the 1794 *Philadelphia Directory and Register* and *The Philadelphia Directory for 1798*, we learn of two yellow fever epidemics and their effects on the city.

Directories also included useful information as well as advertisements. *The Baltimore Directory for 1802* helpfully printed an abstract of the revenue laws, import duties, places of public worship, listings of public buildings, and also advertised such things as land and water stages, pleasure gardens, and baths.

The *Miners and Business Men's Directory for . . . 1856 . . . Citizens*

Book-Binding, Ruling and Lithographing
SAN FRANCISCO
F. S. FRENCH YOKE SHIRTS
CITY OF NEW YORK.
FIRE DISTRICTS
THE BOSTON DIRECTORY.

of Tuolomne (County), provides data about mining communities in early California. Names of the camps, their laws, and a history of each settlement are given, along with listings of the residents of each camp, their occupations, and previous places of residence.

Because of their diverse content, directories have become a valuable tool for researchers in many fields. The Society's collection of Worcester directories extends into the twentieth century. Some Canadian directories are also included in the collection. There are also yearbooks of trades as well as railroad directories in separate collections, each having its own checklist located in the reading room.

The American Antiquarian Society's collection of over 6,500 directories published before 1877 is one of the largest in the United States. In 1961, AAS published Dorothea N. Spear's *Bibliography of American Directories Through 1860*, an important contribution to this subject. The Society has most of the directories listed in this bibliography and continues to acquire important directories when available. Directories listed in the Spear bibliography are available on microfiche, produced by Research Publications Inc., and are shelved in the Society's microform reading room. AAS directories printed through 1820 are fully catalogued in the Imprints Catalogue and are shelved with the Dated Books and Dated Pamphlets collections. Directories published in 1821 or later are uncatalogued but are alphabetically listed by name of town and city in a continually updated checklist. [J.M.P.]

Maps and Atlases

The collection of maps focuses on the United States, although there are maps of all parts of the world that were printed in the United States. The size of the collection is about 10,000 items. The strengths of the collection are maps of New England of the eighteenth and nineteenth centuries and, more specifically, maps of Massachusetts. In 1910, the New England Historic Genealogical Society placed its collection of maps on deposit at AAS; this agreement was recently renewed.

No major acquisitions comparable in size to the collections of John W. Farwell or Thomas W. Streeter have arrived at the Society since 1958, and purchases of cartographic materials each year are restricted to a few items. Gifts are always welcome. Although the geographical arrangement and index of the map collection are adequate for answering reference questions posed by local historians and genealogists, it is almost impossible to respond to queries posed by cartographic historians with respect to cartographers or to techniques used in the compilation of a map. There are partial indices to the collection by engraver and by lithographer. Easy access to those printed in America is made possible by the bibliography compiled by James Clement Wheat and Christian F. Brun, *Maps and Charts Published in America Before 1800* (New Haven & London, 1969). This was reprinted in 1985. Kenneth Nebenzahl's *Bibliography of Printed Plans of the American Revolution* (Chicago, 1975) also reflects the holdings of AAS. These bibliographies are useful to anyone using the collection because they provide access points not available in other finding aids at the Society.

The atlas collection is housed in several different locations. Those printed in the United States before 1821 are catalogued with the other American imprints; most of them are shelved in the Dated Books collection. Access to all of them can be achieved by the use of the genre heading 'Atlases' in the Imprints Catalogue. Those that are too large for shelving in the bookstacks are housed in the graphic arts department. The maps in most of these atlases were described as part of the Catalogue of American Engravings Project, although it was subsequently decided to exclude maps from the catalogue because the description of variant states was too complex. Information from these entries has been transferred to cards that have been interfiled with the cards describing the map collection.

Later atlases are filed in three sections. The first (about 100 volumes) contains atlases that are quarto in size; most of these are school atlases, arranged chronologically by publication date. A second section (230 volumes) contains atlases of the United

States. National atlases are shelved first, then the arrangement is by state, county, and city. The final section is for world atlases (about 80 volumes). These are filed chronologically. Modern reprints or historical atlases are listed in the General Catalogue.

[G.B.B.]

AMERICAN INSTITUTIONS

General Institutions

Alexis de Tocqueville noted the rise of voluntary associations in the United States during the Jacksonian period, and the wealth of materials about such groups in the Society's collections reflects this 'institutionalization' of America. Approximately 23,200 items, most of which are pamphlets published by various kinds of organizations during the years 1821–76, constitute a collection of institutional reports, proceedings, constitutions, bylaws, historical sketches, membership lists, prospectuses, and promotional literature. The institutions range in nature from businesses to voluntary associations. Banks, churches, insurance houses, social clubs, hospitals for the physically and mentally ill, missionary groups, racetracks, temperance associations, museums, and masonic groups are a sampling of the establishments represented. (Only auction houses, canal corporations, railroads, educational institutions, and book, art, and manuscript dealers, all of which are housed in separate collections, are wholly excluded.)

The collection is arranged as follows: institutions that operated or operate on a nationwide basis are shelved alphabetically by title of organization; statewide (or 'general') organizations are arranged alphabetically by title within separate series for each state; and organizations that claimed or claim only particular cities, towns, or counties as their province are filed under town or county, then title, in 'local' boxes that follow each state's 'general' institutions. At this writing, checklists provide access to all the national institutions and to the general and local institutions of six states: Connecticut, Maine, Massachusetts, New Hampshire, Rhode Island, and Vermont. The staff is able to search through the remainder of the collection by name of institution on demand. It should be noted that the presence in this uncatalogued collection of any given institution's publications does not preclude the presence of other publications by the same institution in various of the Society's catalogued collections. [K.A.]

Canals and Railroads

The Thomas W. Streeter Collection on transportation forms the core of the American Antiquarian Society's collection of materials on canals and railroads. This outstanding collection was given to AAS by Thomas Winthrop Streeter, past president of the Society and preeminent collector of railroadiana. The collection now numbers over 6,000 pieces.

The first dated entry in Thomas Richard Thomson's *Check List of Publications on American Railroads before 1841* (New York, 1942) is for Oliver Evans's *The Abortion of the Young Steam Engineers Guide* (Philadelphia, 1805). A copy of this volume survives at AAS with many other reports and articles. In succeeding years, there were many reports and articles on the possibilities for public roads, canals, and railroads. Some four thousand miles of canals were built between 1815 and 1860, chiefly in New York, Pennsylvania, and Ohio, but strong regional rivalries prevented the development of a national canal system, with the Erie Canal the most successful.

Not until successful trials of the steam locomotive were reported from England was any concentrated effort made to establish railroads in America. The first transcontinental railroad was begun in 1863 and completed in just over five years by the Union Pacific and the Central Pacific railroad companies. A Union Pacific baggage car carried a printing press on which Legh Freeman published the *Frontier Index* at twenty-five different locations along the route. The issues from Julesburg, Colorado, and Fort Saunders,

Wyoming, are in the Society's newspaper collection. A group of Bostonians made the first transcontinental trip to San Francisco in 1870 and chronicled the events and scenes in *The Transcontinental*, published in twelve numbers, the first at Niagara Falls on May 24, the last on return to Boston on July 4. A complete set is in the Society's collection of periodicals.

Many of the western railroad companies received large land grants along the right-of-way. As the Society's collection reveals, a flourishing business found land agents with promotional tracts in Europe as well as in the eastern United States. Graft was involved as well, and assorted pamphlets tell of embezzlements, railroad rings, and the Credit Mobilier building company scandal.

Growth of the railroads led to a large service industry, with trade catalogues for engines, cars, and parts, as well as broadsides and brochures for freight rates and regulations, construction specifications, and operator manuals such as *The Road-Master's Assistant and Section Master's Guide*, by William S. Huntington. Many issues of *Appletons' Railway & Steam Navigation Guide* and other regional and national timetables and guides are in the collection, as well as business directories issued by the railroads that provided information on the towns and businesses along the routes. The Society's collection includes material on such groups as the Boston Association of Locomotive Engineers, the Master Car Builders' Association, the Eastern Railroad Association, the National Narrow Gauge Railway Convention of 1872, and the New York Sabbath Committee (whose concern was that the railroads end their operations on the Sabbath).

Access to the collection requires persistence. Much of the canal material is catalogued, as is a large section of secondary works on both canals and railroads. A smaller percentage of primary railroad pieces are catalogued, including a group of official reports concerning western surveys that were part of the collection of western Americana given the Society by Donald McKay Frost. Because so many early references and reports are to be found in uncatalogued government publications, researchers are urged to read the section on government documents in this guidebook and to consult the finding aids.

The uncatalogued collection of canal and railroad materials includes one box of general canal material, filed in chronological order by date of publication, and ten boxes dealing with specific canals, arranged in alphabetical order by name of the canal. In similar order are twelve boxes of general railroad material, over ninety boxes for specific railroads and two scrapbooks of railroad passes. The general boxes contain items such as speeches, trade and commerce reports, route proposals and arguments, and convention proceedings, while the canal and railroad company boxes consist mainly of corporate and engineers' reports. A few foreign railroad companies are included. In addition, there are eighteen boxes of regional and national timetables and guides, and over fifty railroad business directories. Checklists are available for these, and there is a listing of all the canals and railroads represented in the boxed material. There is also an incomplete checklist of articles on the Blackstone Canal that appeared in Massachusetts and Rhode Island newspapers from 1822 to 1837.

Other canal and railroad material is found in many broadside advertisements, lithographs, maps and charts, railroad bonds, passes and tickets, and sheet music, located in the graphic arts department. Related material may be found in the Society's collections of Newspapers and Serials, Trade Catalogues, American Institutions, Stereographs, and Miscellaneous Pamphlets.

Thomson's railroad bibliography, Evald Rink's *Technical Americana: A Checklist of Technical Publications Printed before 1831* (New York, 1981), and Lawrence B. Romaine's *A Guide To American Trade Catalogs 1744–1900* (New York, 1960), are annotated for AAS holdings, and the Society subscribes to the *Bulletin* of the Railway and Locomotive Historical Society.

With over a thousand canals and railroads represented, the collection offers many opportunities for the scholar. Its diversity and depth illustrates not only the growth of a national transportation system but also chronicles the paths of continental migration and the great influence of canals and railroads upon the people and institutions of the nineteenth century. [C.A.A.]

RAIL ROAD MARCH
Piano Forte
MAESTOSO
HANNIBAL & ST. JOSEPH
RAILROAD GUIDE
WESTERN IOWA,
KANSAS, NEBRASKA
GOLD MINES
ROCKY MOUNTAINS, OREGON, UTAH,
AND CALIFORNIA.
J. Q. A. BEAN.
PROCEEDI
Stockholders
RALEIGH & GASTON RAIL ROAD COMPANY.
RALEIGH:
GRAND CANAL CELEBRATION.
VIEW OF THE FLEET PREPARING TO FORM IN LINE.

Educational Institutions

The American Antiquarian Society has excellent collections of official publications of academic institutions, numbering almost 30,000 items for American colleges and universities and almost 5,000 for schools and academies. There is a name checklist for them in the Society's reading room. The collection of materials concerning fraternities is very small, and there is no bibliographic access to it. Materials in all these collections include catalogues, president's reports, alumni catalogues, histories, obituaries, class reports, registers, and calendars pertaining to educational institutions of the seventeenth, eighteenth, and nineteenth centuries.

[C.A.A. and N.H.B.]

Learned Societies

The publications of American and foreign learned societies, past and present, form a collection closely related to the General Institutions collection. The societies included in the collection range from the Academy of Natural Sciences of Philadelphia to the Zamorano Club of Los Angeles. The collection includes approximately 6,600 volumes (in bound or pamphlet form) of transactions, proceedings, annual reports, and miscellaneous publications, some of which are catalogued in the Society's General Catalogue. This collection is also an archive for all of the Society's own publications; the *Transactions*, the *Proceedings*, as well as any books and pamphlets published by AAS are shelved here. The United States learned societies (for which there is a checklist by name of the 195 organizations represented in the collection) are arranged on the shelves alphabetically by name of institution. Foreign learned societies, to which there is at present no bibliographical access, are grouped by country and arranged alphabetically by name within each group. [K.A.]

CANADIANA

The American Antiquarian Society contains an outstanding collection of Canadiana. This should not be surprising in light of the geographical, cultural, and historical links between the United States and Canada. The earliest bylaws of the Society outlined a broad goal to collect and preserve materials from the Western hemisphere. That range has been greatly modified since 1812, but the Society still continues to acquire printed materials relating to the history of New France and British North America from the period of European settlement through 1876.

Journals of early discovery and exploration, nineteenth-century guidebooks, illustrated reports of expeditions, biographies, essays in Canadian folklore and literature, and federal and provincial government documents are a part of the diverse Canadiana collection. The numerous general histories range from De Charlevoix's *Histoire et Description Generale de la Nouvelle France* (Paris, 1744) and the comprehensive twenty-three-volume Edinburgh edition of *Canada and Its Provinces* (Toronto, 1914) to recently published histories focusing on the Canadian experience through 1876.

The Society holds important documentary series such as *Publications of the Canadian Archives*, *Jesuit Relations and Allied Documents*, and the *Publications of the Champlain Society*. All are rich in historical detail and cover a wide range of subjects and major events in Canada's history. The Society has numerous works concerning the Hudson's Bay Company including the *Publications of the Hudson's Bay Record Society*. The Society also holds collections of documents published by the Champlain Society, including the multivolume *Hudson's Bay Company Series* and the *Works of Samuel de Champlain*. The Champlain Society's *The Rebellion of 1837 in Upper Canada* focuses on one of the central events of Ontario's early history and AAS holds several other works about that armed insurrection in Upper and Lower Canada. Among the primary sources on the patriot uprising at AAS are several rare personal narratives of political prisoners transported to the British penal colony in Van Dieman's Land following the rebellion of 1837–38. The collection includes material on the provincial and municipal level as well. In addition to monographs, there are important series from Canadian historical societies and provincial archives. While all regions of Canada are represented, the largest number of studies are for the provinces of Ontario and Quebec.

Canadiana is also represented in several other collections at AAS. There is a noteworthy collection of French-Canadian genealogies as well as family histories relevant to Loyalist studies. An excellent collection of maps and lithographs of Canadian cities is found in the graphic arts department. Some Canadian imprints issued through 1830 are located in the Dated Books, Dated Pamphlets, Broadsides, and Reserve collections. The Society has a representative run of early Canadian newspapers, nineteenth-century almanacs, and scholarly periodicals such as *BC Studies*, *Acadiensis*, and the *Canadian Historical Review*. Government documents are located in both the catalogued Canadiana collection and the uncatalogued foreign documents collection. These official publications reflect the subject interests of the Society and range from a selection of Parliamentary sessional papers and early provincial statutes to committee reports on railroads and canals, annual reports of the Department of Mines, and occasional papers in archaeology and history from Canadian Historic Sites, a division of Parks Canada.

The General Catalogue provides primary access to the Canadiana collection by author and subject headings. The Imprints Catalogue includes some works issued from the Canadian press through 1830 as well as genealogies. To reflect additions to the Society's collection, three of the standard Canadiana bibliographies are annotated by the acquisitions staff. These are M. Tremaine's *A Bibliography of Canadian Imprints 1751–1800* (Toronto, 1951), the Toronto Public Library's *A Bibliography of Canadiana* (Toronto, 1934), and R. E. Watters's *A Checklist of Canadian Literature and Background Materials 1628–1960* (Toronto, 1972).

The Canadiana collection at the Society was greatly enriched in 1964 when Dr. Gabriel Nadeau donated 1,363 works amounting to more than 3,000 volumes on the history and culture of French Canada. During his long career in public health medicine, Dr. Nadeau was also a novelist, historian, literary critic, and archivist. Through the generous gift of his impressive Canadiana library, the Society acquired many notable works, including the first report of the Societé des Bon Livres of Quebec (1843), M. J. and G. Ahern's *Notes pour Servir a l'Histoire de la Médicin dans le Bas-Canada* (Quebec, 1923), J. E. Roy's *Histoire de la Seigneurie de Lauzon* (Levis, 1899–1902), and C. Tanguay's *Dictionnaire Genealogique des Familles Canadiennes* (Quebec, 1871–80). The Society continues to add primary and secondary works to its Canadiana collection, which serves as an important source of historical material for research in Canadian studies. [J.D.C.]

LATIN AMERICANA

At its inception it was the aspiration of the American Antiquarian Society to be 'American' in the broadest sense of the word. In the original bylaws Isaiah Thomas stated that it was the objective of the Society to collect 'books of every description, including pamphlets and magazines, especially those which were early printed either in South or North America.' Collecting Spanish Americana remained a strong interest of the Society throughout the nineteenth century and into the early decades of the twentieth century. In 1868, the Isaac and Edward L. Davis Fund was established for the purchase of materials relating 'to that portion of North America lying south of the United States.' Subsequently, the scope of the fund was broadened and its income used for the purchase of works relating to all of Latin America. Stephen Salisbury, Jr., was interested in Central America and was instrumental in helping the Society develop its collection in the area of Central American anthropology and archaeology. During the early years of Clarence S. Brigham's tenure as librarian of the Society, the collection expanded, particularly in the area of early printing history. It was at this time that all of the bibliographical works

of José Toribio Medina were acquired, as well as a large collection of Mexican almanacs and imprints. In addition, Henry R. Wagner gave the Society a considerable number of important books on Latin America.

In more recent years, it became clear that AAS could and need not cover the entire span of the American experience at a level useful for serious research. Therefore, in 1968 the AAS Council reaffirmed practices that had been followed for a number of years and accepted the proposition that the range of collecting would include the former French and English parts of North America from the period of settlement by Europeans through 1876. As a result of this decision, the Latin American collection was deliberately and extensively weeded, with many European and South American imprints going to Brown University. Since then, only occasional additions have been made to the collection, with acquisitions restricted to West Indian imprints and to books and pamphlets dealing with Central and South America and the West Indies (generally relating to history, relations with the United States, or description and travel) that were printed in the United States before 1877. Secondary historical works, especially those concerned with the social, political, and economic history of the West Indies, and those concerned with the relations between the United States and Mexico during the period of westward expansion are added with greater regularity.

Most of Latin Americana can be accessed through the General Catalogue, and each country or region is given its own number in the AAS classification scheme (H120 Andes—H880 Windward Islands). Almanacs (numbering over 500 issues) and newspapers are the major exceptions; they are shelved with other examples of their genre. [D.N.O.]

FOREIGN HISTORY AND TRAVEL

As suggested by titles included in booksellers' and library catalogues, works of foreign history and travel were very popular with the American reading public during the eighteenth and nineteenth centuries. While the American Antiquarian Society does not attempt to be comprehensive in this area and does not regularly add to this classification, it does house a well-established collection that includes most of the standard works of history reprinted in this country prior to the Civil War, among which are numerous editions of Rollin's *Ancient History, Lectures on Modern History* by William Smyth, as well as copies of many of the original editions imported from Europe, among which is the forty-four-volume set of *The Modern Part of an Universal History, from the Earliest Account of Time* (London, 1759–66). The collection includes a preliminary section encompassing general works of history and travel as well as secondary sources on historiography. Following this are separate classes for ancient history, each of the European countries, and other countries and regions.

Because of the close ties between English and American history, the collection is particularly strong in the area of English history and description. Students of British colonial history can find at AAS such primary source materials as the *Calendar of State Papers, Colonial Series, America and the West Indies*, (London, 1893–1969) and the *Journal of the Commissioners for Trade & Plantations* (London, 1920–38). Other useful sources of British history and description are the many parish registers and travelers' guide-books.

The China trade aroused considerable interest in Chinese history and culture. One unusual item in the AAS collection relating to China is *A Guide to, or Descriptive Catalogue of the Chinese Museum, in the Marlboro' Chapel, Boston*, by John R. Peters, Jr. (Boston, 1845). During the antebellum period, the movement for the colonization of Liberia gave rise to a body of literature describing that country and Africa in general. In 1852, for example, the American Colonization Society published *Information about Going to Liberia, with Things Every Emigrant Ought to Know.*

The missionary movement was also responsible for producing a variety of works on some of the more remote regions of the world. In 1836, the second edition of David Abeel's *Journal of a Residence in China* was published in New York. And during the

1870s Samuel Colcord Bartlett wrote a series of historical and descriptive sketches of the missions under the control of the American Board of Commissioners for Foreign Missions, including those in Africa, Ceylon and India, and Turkey. Like the missionaries, British officers wrote accounts of their experiences and many of these narratives were reprinted in the United States. Examples included in the Society's collections are *Cabool, a Personal Narrative of a Journey to, and Residence in That City in the Years 1836, 7, and 8*, by Lieut. Col. Sir Alexander Burnes (Philadelphia, 1843) and *The Opium War, Being Recollections of Service in China*, by Capt. Arthur Cunynghame (Philadelphia, 1845).

Of particular interest, and actively collected by the Society, are the accounts and journals of American travelers abroad. Recent acquisitions of this genre include *Thirty Days Over the Sea: A Holiday Ramble in the Old World* (Cincinnati, 1873), by J.S. Cantwell and *Winter and Spring on the Shores of the Mediterranean* (New York, 1870), by J. Henry Bennet. Other unusual examples in the collection are *Running Sketches of Men and Places*, by George Copway, chief of the Ojibway Nation (New York, 1851), David F. Dorr's *A Colored Man Around the World* (Cleveland, 1858), and *Correspondence of Palestine Tourists* (Salt Lake City, 1875), a series of letters describing the pilgrimage of a party of influential Mormons, among whom were George A. Smith and Lorenzo Snow.

Recently, the Society acquired a series of catalogues issued by the governments of Australia, Austria, Belgium, France, Germany, Great Britain, the Netherlands, Norway, Portugal, Sweden, and Switzerland that describe their exhibits at the Centennial Exhibition in Philadelphia in 1876. Although these catalogues are classified and shelved with other materials relating to that exhibition, rather than with the history and description of the countries they represent, they are a wonderful source of historical and descriptive information.

About 5,500 titles are included in the Society's collection of foreign history and travel. Additional and related materials are found in the Dated Books and Dated Pamphlets, Reserve, and the Latin-American and the Canadian collections. Access is through the General Catalogue, with subject entries for geographic place names followed by the subdivisions 'History' and 'Description and travel.' [D.N.O.]

THEOLOGY AND RELIGION

Religion

X is the library's major classification for books on religion. The collection covers the whole range of American religious experience, representing all religious denominations and sects, Christian and non-Christian. Among the many subjects classified within X are works on Adventists, Baptists, Congregationalists, deists, Episcopalians, Jews, Mennonites, spiritualists, and transcendentalists.

The collection is approximately 6,500 volumes strong. It consists primarily of books rather than pamphlets, and does not include, for example, the fifteen to twenty thousand individual sermons housed in the Miscellaneous Pamphlets collection. All books classified in X are catalogued in the General Catalogue. Books acquired in recent years are fully catalogued by author, title, and subject, but those acquired in earlier decades are less fully catalogued.

Three broad categories of religious material are catalogued under this classification. In the first category are American editions on theology, devotional literature, doctrinal works, and religious controversy for the period 1821–76. The Society collects comprehensively in this area. The second category includes non-American editions of religious works with strong relevance to American religious history. Particularly notable in this category are the seventeenth-century English editions of works by ministers and theologians like Ames, Shepard, and Davenport. Although the Society does not now actively collect such editions, this part of the collection is useful to the student of colonial religion. Many of our copies were originally owned by important American ministers. The third group of religious materials in the

Society's collection comprises secondary works of any publication date that treat the history of religious ideas, denominations, and movements in America through 1876.

Several other catalogued collections contain materials useful to the student of American religious history. All American editions (on any subject) published before 1821 are fully catalogued in the Imprints Catalogue. Works treating the history of religion in particular states and cities are part of the Local History collection; these include histories and manuals of individual churches and are all catalogued in the General Catalogue. Devotional works for children, catechisms, and many of the publications of the Sunday school and religious tract societies are included in the American Children's Books collection.

Among the uncatalogued collections, note should be made of General Institutions, the Mather Family Library, Miscellaneous Pamphlets, and that of local and national institutions (for annual reports and similar official publications of such bodies as the American Sunday School Union and the Methodist Episcopal Church). [R.C.F.]

The Mather Family Library

More than 1,500 printed books that once belonged to Richard, Increase, Cotton, and Samuel Mather, their families, colleagues, and correspondents, constitute the American Antiquarian Society's Mather Family Library. This collection, the largest extant portion of colonial New England's most important library, has not been studied at first hand by any of the scholars who in this century have written major or minor works on the various Mathers. The library remains, after 173 years, nearly an untapped source for the intellectual history of a significant American family.

Isaiah Thomas acquired the bulk of the collection for the Society from Hannah Mather Crocker (Samuel's daughter) in 1814. (A small number of additions have been made by gift and purchase since that date.) During the winter of 1814–15, Thomas sifted, conserved, and catalogued the library, and appraised its every volume. He wrote out the catalogue in his own hand on twenty-eight sheets, folded to form three folio quires. On the pages of the first quire, Thomas listed the smallest of the books, the octavos, duodecimos, and 32mos; on the second, he listed the larger books, folios and quartos, and the family's manuscripts; and on the third quire, the contents of twenty-five volumes of bound tracts. At the top of the first page of each quire, he inscribed a heading that summarized its contents, thus:

1. 'Catalogue of Dr. Cotton Mather's Library, purchased by Isaiah Thomas and by him given to the American Antiquarian Society.'

2. 'Remains of Mather's Library Folio & 4to. Purchased by I. Thomas and by him presented to the American Antiquarian Society.'

3. '17th Century English Tracts printed from 1590 to 1730, and bound in 25 vols. and numbered, 1 to 25. Principally on the Religious and Political Disputes of the Time.'

Within each of these sections, Thomas arranged the Mather books in roughly alphabetical order, generally according to author or title but often according to genre (e.g. sermons) or a capriciously constructed short title. Within the folio & quarto section, there are separate alphabetical series for each format. Thus, Thomas's manuscript catalogue comprises three sections, four alphabetical series, as well as a list of manuscripts.

Readers will perhaps excuse such detail if they compare this account with the list of books published under the heading, incorrectly transcribing Thomas's words, 'Catalogue of Dr. Cotton Mather's Library purchased by Isaiah Thomas and given by him to the American Antiquarian Society,' found at the end of J. H. Tuttle's 'The Libraries of the Mathers' in the *Proceedings of the American Antiquarian Society* 20 (1910): 269–356. Here an assistant to Tuttle silently neatened Thomas's catalogue into a single alphabetical list, deleting, in the process, more than 500 titles. Gone, for instance, are Thomas's entries for *The Anatomy of Melancholy*, Lancelot Andrewes's *The Morall Law Expounded*, Ainsworth's

Annotations upon the Five Books of Moses, the Book of Psalms, and the Song of Songs (1639), among others.

Some years after Tuttle glanced over these books, the Society's project to recatalogue American imprints got underway. To facilitate it, tract and sermon volumes in the Mather Family Library that contained American imprints were disbound, their American imprints segregated, and all titles separately rebound. Regrettable from the standpoint of association value, this work was done thoroughly. Only two items were passed over: John Higginson's four-leaf *Direction for a Public Profession in the Church Assembly* (Cambridge, Mass., 1665) and the broadside *Catalogus, eorum qui in Collegio Harvardino* (Boston, 1700). When Thomas's catalogue does not specify the original contents of collections broken up for this project, they can be reconstructed with the help of a nineteenth-century shelflist.

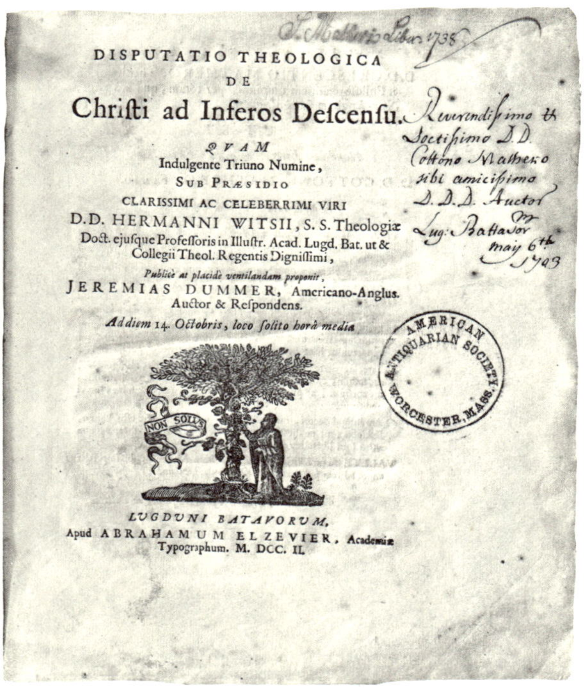
DISPUTATIO THEOLOGICA
DE
Christi ad Inferos Descensu.
QUAM
Indulgente Triuno Numine,
SUB PRÆSIDIO
CLARISSIMI AC CELEBERRIMI VIRI
D.D. HERMANNI WITSII, S. S. Theologiæ
Doct. ejusque Professoris in Illustr. Acad. Lugd. Bat. ut &
Collegii Theol. Regentis Dignissimi,
Publicè at placidè ventilandam proponit,
JEREMIAS DUMMER, Americano-Anglus.
Auctor & Respondens.
Ad diem 14. Octobris, loco solito horâ media

NON SOLUS

LUGDUNI BATAVORUM,
Apud ABRAHAMUM ELZEVIER, Academiæ
Typographum. M. DCC. IL.

Each volume in the Mather Family Library contains some marginalia; at the very least, each bears Thomas's 1814 appraisal. Signatures, dates, jotted purchase information, or gift inscriptions adorn a high percentage of the books. Claudius Gilbert's *Vindication of the Magistrates Power in Religious Matters* (London, 1657), for example, is inscribed 'John Wilsons booke, ex dono authoris,' and 'Richard Mathers booke, ex dono Reverendi John Wilson.' *Anti Baal-Berith* is inscribed 'Tho: Shepard's: ye gift of Col: Tho: Temple. July. 1661.' Less dramatic inscriptions than these are of course equally pertinent to the study of book circulation in colonial New England. Are the upside-down figures on the flyleaf of Increase's copy of Rivet's *Theologicae & Scholasticae Exercitationes cxc in Genesin* (Leyden, 1633) stigmata earned at Samuel Gerrish's auction of September 2, 1718? Scholars studying the writings of the various Mathers will, sooner or later, find the textual marginalia, underscoring, doodles, and inserted indexes of interest. Read in conjunction with the Mathers' own published writings, with the Mather manuscripts held at AAS, and with the several manuscript booklists compiled at different times by Increase and Cotton Mather, even the shortest dash penciled in a margin points its maker as well as the passage. The marginalia in *A Letter of Advice to the Churches of the Non-Conformists* (London, 1700), for instance, are Cotton Mather's own. It is worth noting that this slender work, in a quite fancy contemporary American binding, escaped the eye of T. J. Holmes, who worked at AAS while compiling his exemplary bibliography of Cotton's writings.

Although the Mather Family Library is still largely uncatalogued, all of its volumes, including those mistakenly omitted from Tuttle's 1910 checklist, are accessible through a main-entry card file marked with shelf numbers. A typescript of Thomas's manuscript catalogue is filed alongside it in the Manuscripts Department. These aids will be of use until a corrective to Tuttle's misleading 1910 list finds its way to print. [K.A.]

The Mather Collection

The American Antiquarian Society holds a rich collection of American and English editions of the writings of Richard Mather of Dorchester, Massachusetts, and of his descendants, from forgotten Warham to never-to-be-forgotten Increase. This collection comprises approximately 515 volumes and is preeminent among the very strong Mather collections at the Boston Public Library, the University of Virginia (which holds the William Gwinn Mather-Tracy W. McGregor Collection), and the Massachusetts Historical Society. These volumes, agreeing in name if not in quality, are segregated from the Society's topical and chronological collections and are shelved as a unit. (An armful of late reprints and frequently used works, such as Cotton's *Diary*, 1911, and Ronald Bosco's edition of *Paterna*, 1976, are shelved in appropriate open access collections; their number is not included in the figure given above.)

Access to the American editions of Mather works printed before 1821 is provided by the Society's Imprints Catalogue and by the North American Imprints Program's records (for American editions printed before 1801), which are available online through RLIN. English editions and works published in America after 1820 are listed in the General Catalogue; English editions printed before 1801 will soon be accessible through the catalogues produced by the Eighteenth-Century Short Title Catalogue Project (ESTC) as well. The Society's holdings are incompletely listed in Wing's *Short-Title Catalogue of Books Printed in England, Scotland, Ireland, Wales, and British America . . . 1641–1700*, rev. ed. (New York, 1972–), Evans's *American Bibliography* (Chicago and Worcester, 1903–59), Shaw and Shoemaker's *American Bibliography* (New York, 1958–), and, of course in Thomas J. Holmes's great bibliographies: *Increase Mather*, 2 vols. (Cleveland, 1931), *Cotton Mather*, 3 vols. (Cambridge, 1940), and *The Minor Mathers* (Cambridge, 1940), which incorporate exhaustive indexes compiled by Joseph Tuckerman Day and George W. Robinson. These bibliographies, together with the Society's Imprints Catalogue and its chronological, geographical, and printers indexes, make the corpus of Mather writings the most diversely accessible of the Society's well-catalogued collections. [K.A.]

Devotional Literature

Devotional literature that was published in this country through 1820 is fully catalogued in the American Antiquarian Society's collections of Dated Books and Dated Pamphlets. The Society also has separate, uncatalogued collections of Bibles, catechisms, and prayer books that were published during the period 1820–76. The Bibles number about 1,200 items, but the other collections are much smaller. There is limited annotation of the Society's Bible holdings in Margaret T. Hills's *The English Bible in America* (New York, 1961). As with the Bibles, prayer books and catechisms are arranged on the shelves by date of publication but there is no bibliographical access to them. [C.A.A. and N.H.B.]

LITERATURE

American Fiction

Because fiction is illustrative of so many aspects of the past, it has become indispensable to our understanding of American social history. In fact, for times preceding sound recordings and film, it is a primary source for much of what we know about the way in which earlier Americans lived. Many novels, then as now, possess minimal intrinsic literary merit, but it can nevertheless be argued that they constitute history in and of themselves, because they mirror contemporary circumstances, customs, events, and popular attitudes.

The collection of fiction at the American Antiquarian Society spans nearly a century—from 1789 (the publication date of the first truly American novel, William Hill Brown's, *The Power of Sympathy*, which was published by Isaiah Thomas) through the

year 1876. Also of significance to researchers are dozens of twentieth-century critical studies that provide glosses and interpretations of hundreds of authors of great or modest talent. Lyle H. Wright's three-volume checklist, *American Fiction* (2d ed., 1969), is an essential guide to this literary genre. Some years ago a count of the entries in volume one, which covers the years 1774–1850, revealed that AAS held 2,110, Yale 1,977, and the Library of Congress and the New York Public Library nearly 1,100 each of 2,772 entries. Of citations in volume two, covering the year 1851–75, the Society holds over 60 percent of the 2,832 entries. Wright's third volume begins just as the Society's chronological upper limit ends, but, still, there is an excellent showing of fiction published in the centennial year 1876.

The scholar studying James Fenimore Cooper will find that AAS holds an unparalleled collection of Cooper's printed works, from his first-published *Precaution* in 1820 through approximately the year 1870. Since 1968 the Society and Clark University have been the co-sponsors of the standard edition of Cooper's works, which proceeds under the editorship of AAS member James

Franklin Beard of Clark's English faculty. In an effort to bring the Cooper collection under bibliographical control, the Society has recatalogued some 1,300 volumes of Cooper's works. These records are now in RLIN.

Another important group of popular fiction at AAS is that of dime novels. These formulaic tales of adventure were issued in numbered series and enjoyed an enormous readership from the last third of the nineteenth century until well into the twentieth. The Society's holdings of the Beadle series up to about 1875 are all but complete and AAS holds strong representations of other less popular imitations of that publisher's offerings. The standard guide to this colorful genre is Albert Johannsen's *House of Beadle and Adams and its Dime and Nickel Novels* (Norman, Okla., 1950). This three-volume work is annotated with AAS holdings, and the Society maintains checklists of the publications of other 'dime novel' firms. [A.C.M.]

First Editions

Although the Society has concentrated its energies in collecting examples of popular fiction rather than the canon of 'Standard American Authors,' the scholar concerned with American literature would be misled if no reference were made to the Society's very substantial collections of first editions of the principal American writers. One of Clarence S. Brigham's proudest achievements was the development of outstanding holdings in this expression of American culture of the eighteenth and nineteenth centuries. The First Editions collection is another example of Brigham's interest in literature. Working in concert with P. K. Foley, a Boston bookseller who compiled his generation's 'definitive' checklist of American literary work, entitled *American Authors, 1795–1895: A Bibliography of First and Notable Editions* (Boston, 1897), Brigham built a superlative collection around the nearly 6,000 works that were written by Foley's 286 authors. To this day, the Society maintains an annotated and greatly expanded version of 'Foley.' Items in the First Editions collection are described in the Imprints Catalogue at AAS and appear in the twenty-volume *Dictionary Catalogue* of AAS holdings published in 1971 by Greenwood.

Jacob Blanck's *Bibliography of American Literature* (New Haven, Conn., 1955–83) comes out of the same tradition as that of Foley (and Joseph Sabin before him). Blanck built on Foley's work as well as that of his own master, Merle Johnson, and used AAS as one of his principal sources for authors from the Federal period to the latter part of the nineteenth century. But AAS does not attempt to rival the complete collections of the standard American authors (except for James Fenimore Cooper) that are held by the Houghton Library at Harvard, the Library of Congress, Waller Barrett's great collections at the University of Virginia, the Beinecke Library at Yale, the Lilly Library at Indiana University, or at the Huntington Library. Nevertheless, the scholar studying American literature will do well to examine the holdings of AAS, for AAS has on its shelves impressions, editions, or even unsuspected titles of national authors that may solve a conundrum for the inquirer. [M.A.McC.]

English and European Authors

During the past twenty-five years AAS has made serious efforts to collect American editions of English and European authors of the later nineteenth century. (One must remember that anything printed before 1821 in what is now the United States was for many years earlier energetically gathered in.) For example, AAS has developed significant collections of Dickens and the other great (and not so great) nineteenth-century English authors, as well as Jules Verne, Russian authors, Paul de Kock, and Alexandre Dumas. The Society has also sought sets of the various nineteenth-century American publishers' series. The earliest American editions of important philosophical, scientific, and political works by European authors also are of great interest and are well represented on the Society's shelves.

[M.A.McC.]

Literary Annuals

Literary annuals arrived in force upon the American publishing scene in the mid-1820s. These books were compilations of fiction and poetry, often of a sentimental or religious nature, frequently illustrated and usually handsomely bound. In writing about the collection of literary annuals at the American Antiquarian Society, Clarence Brigham focused on the very important contributions of American authors to this genre of American imprints. Among the writers who contributed their poems, stories, and essays were William Cullen Bryant, Lydia Maria Child, Ralph Waldo Emerson, Edward Everett, William Lloyd Garrison, Nathaniel Hawthorne, Oliver Wendell Holmes, Washington Irving, Henry Wadsworth Longfellow, James Russell Lowell, Frances S. Osgood, Theodore Parker, Edgar Allan Poe, Lydia H. Sigourney, John Greenleaf Whittier, Nathaniel Parker Willis, and others.

It is surprising, given his interest in American graphic art, that Brigham bypassed the illustrations in the literary annuals. In general, these are steel engravings that reproduce paintings, drawings, and watercolors by America's artists including Thomas Doughty, Alvan Fisher, William Sidney Mount, George Loring Brown, and William Guy Wall. A significant number of the illustrations reproduced works by European artists of the Renaissance and later eras and served to make the works of these artists known to a portion of the American public.

The Society's collection numbers approximately 1,400 items, including reissues, variant editions, but not including duplicate copies or copies of a given edition with minor variations. There is a short-title list of these annuals that was compiled in 1983. Titles printed before 1831 are written on yellow cards to facilitate the retrieval of these earlier imprints. The Society's collection contains several English and French imprints, although the vast majority were, of course, printed in the United States.

Other means of access to the collection include Frederick Winthrop Faxon's *Literary Annuals and Gift Books* (Boston, 1912) and Ralph Thompson's *American Literary Annuals and Gift Books, 1825–1865* (New York, 1936). Both are annotated to reflect the Society's holdings. Another finding aid is *Indices to American Literary Annuals and Gift Books, 1825–1865*, compiled by E. Bruce Kirkham and John W. Fink (New Haven, 1975). The 469 titles in this work are keyed to the microfilm edition of the annuals listed by Thompson and reproduced in 1966 by Research Publications, Inc. Part 1 provides basic information on each of the titles, including editor, title, place of publication, publisher, date, table of contents (with the names of the authors, if provided in the volume), and a list of illustrations with the names of artists and engravers (again, if found in the volume). Part 2 contains the several indices for editors, publishers, cities and states of publication, stereotypers and printers, titles of poems and stories, authors, titles of illustrations, painters, and engravers. No attempt was made to attribute literary or artistic works; the indices are based on text within each volume. The indices are research tools for both art historians and literary historians. The Society does not own the microfilm edition of the literary annuals, but it has copies of most, if not all, the titles indexed by Kirkham and Fink.

Unlike some of the special collections at AAS that came as the gift of just one or two donors, this collection has been acquired during the present century from a host of booksellers. A few titles came to AAS as gifts of donors such as Stephen Salisbury III and Charles Henry Taylor. The most recent acquisitions have been part of the collection of American bindings bequeathed to AAS by Michael Papantonio. This large and important acquisition was reported in volume 93 of the *Proceedings* (1983) and in *Early American Bookbindings from the Collection of Michael Papantonio* (Worcester, 1986), which contains indices by binders, previous owners, and an index of main entries. [G.B.B.]

Plays

Dramatic works published through 1820 in this country are catalogued in the American Antiquarian Society's collection of

Dated Books and Dated Pamphlets. These holdings are annotated in Frank Pierce Hill's *American Plays Printed 1714–1830* (New York, 1970) and Oscar Wegelin's *Early American Plays 1714–1830* (New York, 1905). The Society also has a collection of almost four hundred plays that were published in America between 1821 and 1876. There is a checklist of these plays, with access by author and title. [C.A.A. and N.H.B.]

CHILDREN'S LITERATURE AND SCHOOL BOOKS

The Children's Books collection at the American Antiquarian Society is, within its scope, the world's finest. Extending from 1700, the date of Cotton Mather's New England supplement to James Janeway's *Token for Children*, to nearly the end of the nineteenth century, the collection chronicles the emergence of an American literature for children. Books written for children can provide important evidence of the forces structuring childhood experience, that is, the ideals and ambitions of the society that wrote and published them. Scholars studying a variety of social issues—immigration, city life, temperance, and antislavery movements—will find rich materials here. The AAS collection of books for children is thus an unparalleled resource for studies in such fields as the history of childhood, child discipline and the education of children, the history of reading and reading communities, and in the history of publishing, printing, and the graphic arts.

American children's books at the Society are listed in different catalogues according to their date of publication. Children's books published before 1821 are part of the Dated Books and Dated Pamphlets collection, while those published from 1821 forward are housed and catalogued as a separate collection.

As d'Alté Welch noted in his *A Bibliography of American Children's Books Printed Prior to 1821* (Worcester, 1972), the Society 'has by far the largest and most interesting collection of children's books seen. . . . [It] contains almost two thirds of all extant American [pre-1821] children's books.' After Welch's untimely death in

1970, our already unrivaled collection was further strengthened by the acquisition of his personal collection.

Most American books for children printed during this pre-1821 period are reprintings of English chapbooks—which included traditional tales like 'Cock Robin' and 'Children in the Wood,' abridgments of literary works like *Robinson Crusoe*, and books of amusement and instruction, particularly those first issued by the London publisher John Newbery. Towards the end of the eighteenth century, such English writers as Maria Edgeworth and Sarah Trimmer began producing a literature more especially designed for children, and their works were quickly reprinted in the United States.

Welch's bibliography covers only the nonpedagogical genres of pre-1821 American juveniles, but the AAS collection is nearly as strong in early primers, textbooks, and catechisms. The Society holds the single strongest collection (about 200 editions) of New England primers printed before 1830, as recorded in Charles Heartman's bibliography *The New England Primer Issued Prior to 1830* (New York, 1934), and nearly half of the non-New England primers recorded by Heartman in *American Primers, Indian Primers, Royal Primers, and Thirty-Seven Other Types of Non-New England Primers Issued Prior to 1830* (Highland Park, N.J., 1935).

American editions of children's books published before 1821 are fully catalogued in the Society's Imprints Catalogue and may be found together under the subject heading 'Children's literature.' American juveniles published before 1801 are also catalogued in machine-readable form as part of the North American Imprints Program. Finally, it should be noted that the Society's copies of important bibliographies like Welch's, A. S. W. Rosenbach's *Early American Children's Books*, and Heartman's *New England* and *Non-New England Primers* are fully annotated to reflect the Society's collections.

Children's books published after 1820 form a separate collection. During this period, American writers and publishers made a deliberate effort to produce a native literature for American children, one suited to instruct the young citizens of a new republic, and this collection is particularly rich in materials for the social historian.

The nonpedagogical portion of this collection is one of the strongest in the nation. It numbers about 9,000 editions and includes fiction, poetry, natural history, travel narratives, Sunday-school tracts, and conduct-of-life manuals. The collection is especially strong in the works of Oliver Optic, Jacob Abbott, and Samuel Goodrich, in chapbooks and toy books, and in the publications of the American Sunday School Union and the American Tract Society. The collection also contains some important boxed sets and cabinet libraries.

One portion of this collection is worth noting separately. The Society holds approximately 1,200 issues of the New York firm of McLoughlin Brothers. McLoughlin publications are particularly well known for their use of colored illustrations, which were hand-stenciled during the firm's early years and printed from etched zinc plates and photo engravings later on. The collection includes many copies from the company archives. The bulk of the collection is the gift of AAS member Herbert H. Hosmer, who had acquired much of it from a former member of the firm. The collection represents a departure from the usual collecting policy of the Society, for the imprints within it range from the 1830s to well into the twentieth century. Current acquisition policy provides for purchase of McLoughlin publications published to about 1890.

The Society also holds significant collections of pedagogical juvenile books, although it does not collect as comprehensively in this area. There are about 1,300 school books, 230 Sabbath-school books and primers, and 320 primers published between 1830 and 1876 in the collection.

Access to the post-1820 juvenile collection is supplied chiefly through a computerized cataloguing project that was begun in 1985. This project, funded by the National Endowment for the Humanities, will proceed in several steps; currently, it covers nonpedagogical books published between 1820 and 1860. In content and format, this database complements and extends the

New-York
CRIES
New-York.
NEW-YORK:
1822.
REUBEN KENT'S
FIRST WINTER IN THE CITY.
"REUBEN KENT AT SCHOOL."
PHILADELPHIA:
AMERICAN SUNDAY-SCHOOL UNION,
CITY SIGHTS
FOR
COUNTRY EYES
The Alphabet.
ABCDEFGHIJK
LMNOPQRST
UVWXYZ.
abcdefghijklmno
pqrstuvwxyz.
ABCDEFGHIJK
LMNOPQRST
UVWXYZ.
abcdefghijklmno
pqrstuvwxyz.
1234567890.
THE
Cries of London.
COOPERSTOWN.

Society's North American Imprints Program. The machine-readable records, which are available in the RLIN database, represent an unusually detailed approach to juvenile materials and are intended to support a wide range of studies, ranging from bibliography to social history.

Several manual files are also available, which cover portions of the collection not yet fully catalogued. There are separate checklists for non-pedagogical works for the years 1821–76, McLoughlin publications, school books (subarranged by subject), primers, and Sabbath-school books and catechisms. Pertinent bibliographies such as the *Bibliography of American Literature* (New Haven, 1955–), Delores B. Jones's *An Oliver Optic Checklist* (Westport, Conn., 1985), and Carl J. Weber's *A Bibliography of Jacob Abbott* (Waterville, Me., 1948) are annotated with AAS holdings.

Finally, it should be noted that the Society's holdings include important supporting materials for studies in the history of childhood and children's literature, including advice manuals for parents and teachers, numerous games (including many manufactured by the McLoughlin Brothers), and the largest single collection of American children's periodicals to 1876 (for which a checklist of titles and holdings is available). The Society also holds a strong collection of juvenile diaries and one of the largest collections of amateur newspapers, many of them written, printed, and distributed by children under eighteen, which provide a rare source of insight into the mentality of historical children. [R.C.F. and S.J.W.G.]

GRAPHIC ARTS

The graphic arts department houses five major collections that are related in medium or format, if not in subject matter: prints, photographs, broadsides and ephemera, maps, and sheet music. In addition, oil paintings, furniture, and other artifacts belonging to the Society are considered part of this collection. The Broadsides, Maps, and Sheet Music collections are described elsewhere in this volume, but most of the other collections will be described here. Unless otherwise noted, collections are not catalogued or indexed; the collections housed in the department are arranged by type or by subject.

Engravings and Lithographs

With the important exception of the lithograph collection, prints are not catalogued, but are arranged by subject and by size. Among the major divisions are the following: United States views, portraits of Americans active through 1900, Hawaiian engravings, engravings printed before 1821, later engravings generally removed from books and periodicals (arranged by the name of the engraver), historical scenes, religious subjects, and English political cartoons.

The emphasis of the collection of American engravings has always been on the works of colonial engravers, particularly those active in New England, more specifically, Boston. The formation of this portion of the collection dates back to the bequest of William Bentley, a Salem clergyman. After his death in 1819, his paintings and prints came to the Society, and many of his prints are on display in Antiquarian Hall. Bentley was particularly interested in portraits, and many of our Peter Pelham mezzotints dating from the first half of the eighteenth century, as well as about a dozen chalk drawings by Samuel Harris (d. 1810), came from his collection. The woodcut portrait of Richard Mather (the earliest American portrait print) also adorned the walls of Bentley's home in Salem. Other early prints, including portraits of the first six presidents of the United States, came from the Society's founder after his death in 1831.

Clarence S. Brigham compiled the definitive work on the engravings of Paul Revere, which the Society published in 1954. During the years leading up to the publication of this monograph, Brigham scouted out impressions of Revere's engravings until the Society had at least one of each, except for the portrait of Jonathan Mayhew. Since then, two engravings by Revere have surfaced;

AAS has an impression of one (a meeting notice for the Relief Fire Society) but not of the other (a billhead for Mr. John Piemont, owned by the town of Danvers, Massachusetts). The Society has holdings almost as strong for other major Boston engravers of the eighteenth century—James Turner, Nathaniel Hurd, and Thomas Johnston. Like Revere, these craftsmen engraved maps, bookplates, currency, membership certificates, book illustrations, and political prints.

For engravings of the nineteenth century, the holdings of the department are not as strong. Political cartoons have always been of interest because of their strong historical associations, and a large number of scarce prints issued before 1820 are in the collection. Portrait prints have not been collected vigorously, but unusual examples, including mezzotints of Gov. Elbridge Gerry and Maj. Gen. Benjamin Lincoln by John Rubens Smith, are at the Society.

Engraved views from the first half of the nineteenth century are among the most beautiful prints issued in the United States. Recent acquisitions to the Society's holdings have included two aquatint views of New York City that were engraved by John Hill in 1823.

Access to American engravings and book illustrations issued before 1821 has been greatly improved during the past few years by a project funded by the National Endowment for the Humanities and the H.W. Wilson Foundation. The Catalogue of American Engravings Project has located and described some 16,400 engravings in the collections of the Society and in libraries and museums across the country. In due course, the publication of this catalogue will enhance research on American engraving and will provide subject access to this vast body of pictorial material.

The lithograph collection was established as a separate entity within the department in 1928, when Charles H. Taylor, publisher of the *Boston Globe* and one of the Society's most generous donors, gave AAS his lithograph collection. He continued to give the Society hundreds of lithographed items each year, especially prints and illustrated books. There are approximately 10,000 separately catalogued lithographs. The catalogue for this collection has been expanded to reflect new acquisitions and segments of other collections that have been integrated into the lithograph collection including panoramic views, portraits, circus posters, and political cartoons. Access to this significant collection is by a card catalogue, which supplies information about the content and iconography of prints, creators, publishers, and copyright holders. This catalogue is an unusual feature of the department; few other institutions have catalogued prints so thoroughly.

Taylor also donated books illustrated with lithographs. These volumes have been indexed by lithographer and by subject. Although additions are not made to this index, it remains a useful reference tool to those interested in lithographs by a specific firm.

In recent years, the purpose of additions to the lithograph collection has been to acquire lithographs with significant historical content, whether political, social, or cultural. Following Taylor's own collecting interests, the Society still tries to obtain the products of the early Boston lithographic presses—those of the Pendletons, Thomas Moore, and John H. Bufford.

The segments of the print collection can be used by scholars who have various interests. Documenting the physical aspects of cities and towns, understanding land use, examining architectural features of specific buildings are possible by examining views of the United States, whether engraved, lithographed, or photographed. Illustrations of familial relationships, rural and urban pastimes, trades and occupations, for example, can be located through the catalogue to the lithograph collection, making that collection useful to historians and picture researchers from many disciplines. Because of the thorough cataloguing of the lithographs, art historians can trace the careers of specific designers and lithographers with ease. [G.B.B.]

WORCESTER HOUSE.
THE WORCESTER TEMPERANCE HOUSE.

Ephemera

The graphic arts department houses many additional collections of what can be termed ephemera. Some of these collections, such as those for bookplates and currency, contain items of great rarity and the collections are among the strongest in the nation. Others are essentially accumulations of materials that librarians and curators at the Society have assembled.

A frequently used collection is that of eighteenth- and nineteenth-century trade cards. Scholars interested in the history of American graphic arts consult them because many are engraved by individuals such as Paul Revere and James Turner. Others refer to the collection because of the interesting design of these small advertisements. Many are illustrated with vignettes of shops or of people working, depictions that are of interest to scholars in several disciplines. There is also a separately filed collection of small trade cards of the latter part of the nineteenth century. These are mainly stock cards that were printed by large commercial printing firms with blanks for the local firm's message. Many of these are printed in color with comical or sentimental illustrations. This collection is largely filed by type of advertiser—groceries, dry goods, medicines, etc.

Related to the trade cards is the collection of watch papers. These ephemeral pieces were inserted into the backs of watches, frequently with the date of the most recent servicing or repair. The most elegant are engraved, and the figure of Father Time appears with regularity. In 1951, the Society published a checklist of the collection compiled by Dorothea N. Spear. A few additions have been made since then, and an annotated copy of the checklist is shelved in the Society's bibliography section.

Colonial currency is a collection of major significance. The collection is mounted in twenty-two notebooks. The bibliography by Eric P. Newman, *The Early Paper Money of America* (Racine, Wis., 1976) is annotated to reflect the Society's holdings. There have been no additions to this collection in recent years because of the high prices of individual pieces of currency.

The Society's game collection is a sizable one of about 170 items that grows as opportunities occur. The games are arranged in two classes: one for instructional games, the other for those that are purely for entertainment. These all tend to be board games or card games, with a few jigsaw puzzles present as well. Some of the latter are geographical; others are picture puzzles. As historians focus on material culture, the formation of attitudes in childhood, and leisure time and its uses, this collection will receive more attention. Games issued by the McLoughlin Brothers are of particular interest because of the relationship of the games to the children's books published by that firm.

The valentine collection features early examples made by Esther Howland of Worcester, as well as examples printed from later in the nineteenth century. It is a representative collection and incorporates some valentines printed in Europe. Related to these is the collection of Louis Prang salesmen's books of that firm's greeting cards. Arranged by function or by holiday, that collection numbers some fifty volumes. Additional lithographs by the Prang firm, which specialized in chromolithography, are catalogued in the lithograph collection. There is also one box of miscellaneous Christmas greetings.

A large collection of rewards of merit is often interesting for the pictorial content of these nineteenth-century ephemeral pieces. They were awarded to students for good behavior or excellence in their studies. The collection is filed by iconography or pictorial content. There is also a group of rewards filed by printers' names.

The graphic arts department includes a number of other minor collections, some of which were described in detail by Brigham. They are national in scope and generally date before 1877. A partial listing of them will indicate to the researcher the riches of this department. In alphabetical order they are: ballots, billheads, bills of lading, calendars, certificates of membership, clipper ship cards, copybook covers, nineteenth-century currency, dance cards, diplomas, labels, lottery tickets, passports, railroad tickets and passes, sentiment cards, silhouettes, stock certificates, telegram forms, tickets to lectures, type-and-banknote-engraving specimen sheets, and wrappers for reams of paper. [G.B.B.]

Portraits and Artifacts

Visitors to Antiquarian Hall and other AAS buildings are generally surprised by the array of antique furniture, artifacts, and portraits. The furniture includes some fine American and English antiques, including John Hancock's double chair, tall clock, and desk that are housed in the Council Room. Another piece favored by American scholars is the high chair used by descendants of Richard Mather, including Cotton Mather. Much of the furniture in the administrative offices has some interest to collectors or admirers of American antiques. An unusual aspect of this collection is the number of working antique clocks, including a Willard banjo clock and a David Wood shelf clock. The most recently acquired example was a gift of Kenneth D. Roberts; it is a clock made in Worcester by Samuel Stowel in 1773 as his apprenticeship piece. The clocks keep our staff and readers alert with hourly chimes. An article on some of the most important pieces in the collection was written by Wendell D. Garrett for *The Magazine Antiques* (March 1970).

Artifacts include two wooden busts of John Winthrop and Voltaire, carved by the Salem architect Samuel McIntire, a vial of tea collected after the Boston Tea Party in 1773, silver from the family of Isaiah Thomas, miscellaneous pieces of silver and china, some of which is used at public functions, and the Emma DeForest Morse collection of American Historical Pottery. This collection of 324 pieces of nineteenth-century Staffordshire pottery, acquired in 1913, illustrates major sites in the United States and commemorates events in this nation's past. It is arranged on walls in the curator's office. Two other artifacts in the collection are Col. William Henshaw's musket dating from the American Revolution and a sword that once belonged to Fitz-John Winthrop, the governor of Connecticut, which has descended through the Winthrop family to the Society.

The collection of oil portraits is displayed throughout Antiquarian Hall. The Society has never actively acquired paintings, except for portraits of presidents of the Society. Over the years, however, a significant collection has been formed. Several portraits of eminent members of the Mather family are present (including two portraits of Cotton Mather and one of Mather Byles, Sr., by Peter Pelham), four portraits by the itinerant artist Ethan Allen Greenwood, and Mather Brown's self-portrait. The collection of oil paintings, as well as the miniatures, the sculpted portrait busts, and a few framed drawings are described in Frederick L. Weis's checklist of the portraits in the library of the American Antiquarian Society printed in volume 56 of the *Proceedings* (1946). Since that time portraits of Timothy Swan, Benjamin Chapin, John Moore, Stephen Peabody, members of the Daniels family, and Samuel Sewall, a painting of the Hongs at Canton, and two American landscapes have entered the collection. An article by Louisa Dresser on the collection was published in *The Magazine Antiques* in November 1969. The Society's portraits are included in two databases, the Catalogue of American Portraits and the Bicentennial Inventory of American Paintings. [G.B.B.]

Photographs

The Society has never had a well-defined collecting policy for photographs, yet a representative collection of nineteenth-century photographs has emerged. Among the treasures of the department, for example, is the daguerreotype of Edgar Allan Poe. There are over one hundred daguerreotype portraits whose subjects have been identified, and at least that many of unidentified subjects. The Society's collection of cased photographs also includes ferrotypes and ambrotypes. Photographic portraits have been accepted into the collection, largely as gifts, for the past century and more. Cabinet-sized photographs were removed from the general portrait file several years ago. An index of that collection identifies the subjects, dimensions, and photographic techniques of these photographs and provides biographical information on the sitters. The smaller carte-de-visite portraits have not been indexed. There is a separate alphabet for portraits of Worcester citizens. There are also separate collections of portraits of actors, actresses, and native Americans. About a dozen family photograph albums are housed in the department as well.

Photographic views are interspersed throughout the collection of United States views. Among the most interesting are a small group of photographs by William H. Jackson of Yellowstone National Park; another group by Andrew Joseph Russell of the construction of the Union Pacific Line through Nebraska and Wyoming in 1868 and 1869; almost sixty photographs of Tuskegee Institute and the surrounding area taken in the late 1800s; and a variety of large format photographic views of Worcester.

One of the special interests of the Society in recent years has been collecting books printed before 1877 that are illustrated with photographs. There is access to these books in the General Catalogue through the heading 'Photographs—specimens,' arranged in order by imprint date. Over 130 titles are listed in this classification including Benjamin Silliman's *The World of Science, Art, and Industry Illustrated from Examples in the New York Exhibition* (1854), *Autograph Etchings by American Artists* (1859), Gustave Brion's *Illustrations to 'Les Miserables' of Victor Hugo* (1863), and books of views of Central Park (1864), Greenwood Cemetery (1868), Yosemite Park (1868), and the Rocky Mountains (1870). The number of titles in this classification increases both by new acquisitions and by the identification of books acquired in the past. [G.B.B.]

Stereographs and Postcards

The Antiquarian Society houses one of the country's largest collections of American stereographs. Stereographs, an early form of three-dimensional photograph, were a major vehicle for popular education and entertainment in the latter part of the nineteenth century. Many nineteenth-century photographers now regarded as fine artists produced significant bodies of work in stereograph form; among these were Timothy O'Sullivan, Carleton Watkins, and Eadweard Muybridge. Stereographs were also used for journalistic reporting on many of the current events of the period: parades, disasters, and political events. The Civil War and the Spanish-American War are also documented on stereocards with textual commentary.

The images in this collection date from the mid-1850s to after the First World War. Most were made in the 1870s and 1880s. The collection includes examples of several photographic technologies, including some rare glass slides of Niagara Falls produced in the 1850s by the Langenheim Brothers. Color and monochrome photolithographs are also found here, although the overwhelming majority of the prints, of course, are albumen.

The collection contains fifty to sixty thousand stereograph cards. Most, forty or fifty thousand, are views of American landscapes and city scenes. The views are arranged by state and by place or city within a state. All regions of the continental United States and Canada are well represented, and Central and South America are represented in smaller numbers.

The remaining 10,000 cards represent a variety of genres. Some depict important historical events such as the Civil War and the Spanish-American War. Others represent types of subjects—sports and games, American Indians, parades and celebrations, and paintings and statuary. Still other cards loosely categorized as 'Modes and manners,' 'Childlife,' and 'Courtship and marriage,' are comic or sentimental genre scenes or narrative sequences. These are a rich source for studies of popular iconography and social attitudes toward blacks, women, the Irish, drunkenness, and other subjects. A complete list of subject categories is available. For a useful survey of the history of stereographs, including assistance in the dating of stereographic cards, the reader may wish to consult William Culp Darrah's *The World of Stereographs* (Gettysburg, ca. 1977), available at the Society.

The Society also holds a collection of approximately 115,000 picture postcards, most of them dating from the 1890s through the 1920s. Again, the majority of these postcards, arranged geographically by state and city, represent American scenes.

[R.C.F.]

Photographs of Sculpture on American Gravestones

An unusual but valuable and growing collection at the American Antiquarian Society is that of the photographs of grave markers. Old burial grounds are treasure houses of early American sculpture and of historical and genealogical information. As Harriet Merrifield Forbes noted in her study on gravestones, 'The colonists used their finest skill and raised their most enduring and characteristic works of art in memento mori.' In recent years, however, these storehouses have been endangered by vandalism, natural erosion (hastened by air pollution), and theft.

The original core of the gravestone photograph collection at the Society consists of a gift given in 1930 by Harriet Merrifield Forbes. This author and pioneering photographer donated five volumes of photographs and hundreds of glass plate negatives. These negatives have since been copied onto modern film. The negatives are indexed in a card catalogue that is arranged by town, with further subdivision by burial ground when needed.

There are approximately 1,300 negatives in this part of the collection, filed by last name of decedent. In addition, there are five boxes of Forbes papers, two of which contain correspondence and historical notes relating to gravestones and gravestone cutters.

The larger and more modern part of the collection is an ongoing gift from Daniel Farber and his wife, Jessie Lie Farber. Beginning in 1967 with a gift of three loose-leaf volumes, it has swelled at this writing to include ten volumes of photographs. Even more useful to the researcher are the approximately 9,000 individual photographs of some 6,500 tombstones. Cataloguing data on the individual photographs is being entered into a database that produces printout indexes of carvers, locations, and decedents. Other data has been entered, but is not currently indexed.

This collection, while not comprehensive, is very strong for central and southern New England, with scattered coverage of Maine, New Hampshire, and Vermont. Also here are photographs of stones from England, the Maritime Provinces of Canada, and the Middle Atlantic and southeastern United States. Most of the markers pictured were made prior to 1800.

The Society also has a growing number of books on the subject, accessible through the General Catalogue under the heading 'Sepulchral monuments.' By building a lasting and careful record of one form of early American art and history, this collection helps to ensure that an invaluable resource will be available for future generations. [M.G.]

Secondary Works in the Arts

The Society's Y classification comprises modern secondary works on American fine, decorative, and applied arts from the colonial era through the American Centennial. The section is divided into sixty-three subdivisions. Within the fine arts, books on painting, sculpture, and the graphic arts are emphasized. The decorative arts subdivision covers a large assortment of subjects: antiques, brassware, candlesticks and candles, chinaware, clocks and watches, copperware, costumes, decoration, decoys, firearms, firebacks, furniture, glassware, ironwork, kitchenware, lamps, needlework, pewter, pottery, powder horns, quilts, rugs, silver, stencils, swords, textiles, tinware, and woodenware. Within the applied arts division are books on bells, buttons, ship models, signs, stoves, tools, toys, and wallpaper. This classification also includes material on printed ephemera: advertising, autograph albums, bookplates, numismatic material, playing cards, political collectibles, postcards, posters, silhouettes, stamps, stereographs, trade cards, and valentines. Although a reader might well expect the Society to collect only secondary works relating to the art holdings of the institution, the collecting policy is, in fact, very broad.

Traditionally, scholarship in the fine arts and in the major decorative arts of furniture, textiles, and silver has been very strong. In recent years, scholarship in the related fields mentioned above has been improving, largely in response to the intense interest in American material culture by individual collectors, museums, and scholars. Although the Society keeps abreast of current scholarship, it does not acquire collectors' guides to prices and similar publications that become dated within months of publication. Exhibition catalogues are collected as they are published, and older exhibition catalogues are purchased if they seem to be significant additions to the literature in the field.

Related to this classification but housed separately is the collection of art auction catalogues, which in themselves are useful research tools. This collection is annotated in *American Art Auction*

Catalogues, 1785–1942: A Union List (New York, 1944), compiled by Harold Lancour. Biographies of artists are housed in the biography section of the library; books and pamphlets related to the fine or decorative arts of a state or smaller geographical unit are usually catalogued in the local history section. It is, therefore, impossible to arrive at a total figure for the Society's holdings related to the arts. As an approximation, however, there are about 4,130 volumes and pamphlets shelved in the art classification alone. [G.B.B.]

MUSIC

Hymnals

The American Antiquarian Society has a vast and diverse collection of early North American hymnals, including a large number of volumes from the preeminent collections formed by Bishop Robert W. Peach and Frank J. Metcalf. Beginning with the first book printed on American soil, *The Bay Psalm Book*, the collection includes compilations of sacred verses and music printed in North America through 1876 and numbers well over 5,000 volumes. Although the term 'hymnal' may convey a nearly static and uninteresting concept to modern ears, the Society's collection of early hymnals proves otherwise.

The earliest hymnals in the collection consist of metrical psalms, the various revisions of which remained popular throughout the eighteenth century. Verses and tunes were often printed separately, but early hymnals are also found with manuscript, printed, or engraved tunes at the end of the work. The separately published tunebooks often include one verse of text, but it is the tune, obviously the more important element, that is labeled. Another noteworthy characteristic of the tune books is the introductory matter, usually a small primer on music, presenting the 'art of singing' or the 'rules of psalmody.' The mid-eighteenth-century hymnals display the new acceptance of original lyrics, incorporating the prolific verses of Isaac Watts and Charles Wes-

ley. The period tunebooks reflect new styles and a wider variety of tunes. It is the nineteenth-century works, however, that include the greatest number and the greatest diversity of hymns; folk hymns, revival hymns, gospel songs, and spirituals are added to the repertoire of North American sacred music, introducing such intriguing hymnal titles as *Seaman's Hymns, Millennial Praises, Hymns for the Ohio Lunatic Asylum*, and *Revival and Camp Meeting Minstrel*. Denominational hymnals in the collection are numerous, but the standard modern-day hymnal of tunes, coupled with many verses of text, were not commonly published until the second half of the nineteenth century. The breadth of the collection continues to grow as unique or unusual hymnals are added. Not to be underestimated in importance, the hymnals collection at the Society reveals both obvious and subtle religious, moral, social, and musical trends of the seventeenth, eighteenth, and nineteenth centuries.

The pre-1821 hymnals are fully catalogued, and are easily accessible via the Imprints Catalogue. Two works helpful in identifying and locating pre-1821 tunebooks are Frank J. Metcalf's *American Psalmody* (New York, 1917) and a checklist of sacred tune books located at the Society, both available at the readers' services desk. Nearly all of the post-1820 hymnals are shelved as a separate collection, although a few find their way into the Miniature Books collection by virtue of their size. A detailed checklist of hymnals is available at the readers' services desk. This catalogue was begun by the late Valmore Goucher as an author-title union list of hymnals, and it continues to be annotated and updated with AAS holdings. Included are all of the English-language hymnals found at the Society, followed by a supplement of post-1820 North American-printed foreign language hymnals, principally in German and French. Another bibliographical tool, to be published by AAS, is *American Sacred Music Imprints, 1698–1810: A Bibliography*, by Allen P. Britton, Irving Lowens, and Richard Crawford. The collection is strongly supported by secondary sources, including John Julian's definitive *The Dictionary of Hymnology* (London, 1907), annotated by Bishop Peach; numerous biographies of hymn writers and composers; *The Bibliography of American Hymnals*, published in microform and compiled by the Hymn Society of America; early and contemporary periodicals; sound recordings; and many denominational and topical bibliographies. [P.M.]

The Sheet Music Collection

The collection of sheet music at the American Antiquarian Society consists of about 60,000 pieces of instrumental, vocal, secular, and religious music by both American and foreign composers that were printed through 1880 (more than 4,100 compositions were printed in the United States before 1826). Although Boston imprints are in the majority, the collection also embraces works published in many other sections of the country, notably New York, Philadelphia, Baltimore, Chicago, New Orleans, and San Francisco.

The music is housed in the graphic arts department, shelved alphabetically by composer. A title index housed with the collection provides a second means of access for researchers. In addition to the title and name of the composer, the index lists the place of publication, publisher, and date. It also indicates any special filing category for retrieval of the music. Both Richard J. Wolfe's bibliography *Secular Music in America, 1801–1825* (New York, 1964) and *A Bibliography of Early Secular Music* [*18th Century*] (Washington, 1945) by Oscar G. T. Sonneck and William T. Upton, are annotated to reflect the AAS holdings and new acquisitions. The numbers assigned by Wolfe are also added to the collection's title index.

The sheet music collection is divided into six categories. The largest contains about 44,400 pieces. As with all the other subdivisions, it is arranged first by composer's name, then alphabetically by title, and, in the case of multiple variant imprints, by place of publication and publisher. Some 9,000 pieces of music with lithographed pictorial covers form a second division. This group is used extensively by researchers. The covers not only give a pictorial dimension to the musical content but also present a social and cultural commentary on the era. Songs were composed to pay tribute to the heroism of fire fighters or to celebrate such important events as the first water piped into New York in 1842. These pieces illustrate such important issues of the day as temperance, slavery, and women's rights. They extolled the pleasures of rowing or bowling, ice cream parlors, or tobacco. Even the appearance of the great comet in 1843 was deemed appropriate for the popular composer and the cover artist. The artists who illustrated the pictorial sheet music included some of the nation's most prominent—Winslow Homer, James McNeill Whistler, David Claypoole Johnston, and Fitz Hugh Lane. Consequently, this category of sheet music is in demand not only by scholars, but also by many publishers who wish to use the illustrations for

Correct Position of the Hands.
POSITION AT THE PIANO.
RICHARDSON'S
NEW METHOD
FOR THE
PIANO-FORTE,
PROGRESSIVE ARRANGEMENT, ADAPTATION AND SIMPLICITY;
RUDIMENTS OF HARMONY AND THOROUGH-BASS.
NATHAN RICHARDSON.
BOSTON.
OLIVER DITSON & CO., 277 WASHINGTON STREET.
THE HERO.
THE LARK;
BY O. B. BROWN.
BOSTON:
Published by G. D. Russell & Company, No. 126 Tremont, opposite Park Street.
BUGLE NOTES
RALLY FOR THE RIGHT
FOR THE TEMPERANCE ARMY
EDITED BY F. SHERWIN & J. N. STEARNS
NEW YORK
NATIONAL TEMPERANCE SOCIETY, 58 READE
WOOD'S MINSTRELS SONGS.

current publications. To aid researchers, there are supplemental indexes to lithographers and titles for this group and a partial index by subject.

The Society also holds about 5,000 pieces of music with engraved covers and another approximately 800 are connected with lyrics written by authors cited in P. K. Foley's bibliography *American Authors 1795–1895* (Boston, 1897). A smaller group of sheet music includes compositions displaying photographic or lithographed portraits on their covers.

A group of about 600 pieces constitutes the 'Worcester Collection.' This category comprises music either composed by a Worcester native, published in Worcester, or celebrating a Worcester subject. This group is the only one that contains imprints extending into the twentieth century. Included are such pieces as the *Rangers Trip to Westborough or Lion Quick Step* by James Hooten, written for the opening of the railroad to Westboro, Massachusetts, on November 15, 1834, and *Good Old Worcester Town*, composed in 1917 by Hamilton B. Wood, a former president of the Worcester County Music Association.

Supplementing the large number of compositions by Stephen C. Foster is a three-volume, privately printed set containing reproductions of all Foster's known works and arrangements, the *Foster Hall Reproductions* (Indianapolis, 1933). The Society also holds a small but significant group of vocal and instrumental compositions that were published as serials. Two volumes of the serial *Musical Journal for the Pianoforte*, first published in 1800 by Benjamin Carr in Philadelphia, are included, as well as a complete four-volume facsimile reprint published in 1972.

Rare items in the Society's pre-1826 group of sheet music include the 1814 Baltimore printing of the second issue of the first edition of *The Star Spangled Banner*, Benjamin Carr's *The Wreath of Roses* (Philadelphia, ca. 1816), and the *Hunters of Kentucky*, composed by William Blondell in support of Andrew Jackson as a presidential candidate in 1824.

The Sheet Music collection is one of the more outstanding collections in the country. The Society continues to augment the collection with as much pre-1826 music as possible. Also of interest is music published through 1880 in the West, Midwest, or South. [A.T.Z.]

Secular Music

The Secular Music collection of the American Antiquarian Society consists of American nonsacred vocal and instrumental music, published as compilations and instruction books. The secular music collection comprises approximately 1,500 volumes and runs from the late eighteenth century through 1890. Several of the works are from the collection of the late Irving Lowens.

The bulk of the collection is formed by four principal types: instructional manuals of music and song books that were intended to be used in schools and Sunday schools (written and compiled by such masters as Lowell Mason and George F. Root); the omnipresent piano music, piano exercises, and piano instructional books that were popular throughout this period; instructional manuals for voice and for other instruments; and opera selections and scores, including both European works and American ballad operas. Also included in the collection are compilations of dance tunes, military tunes, college songs, minstrel tunes, national songs, and temperance songs, all including musical notation.

Clearly, the published works reflect the musical taste of Americans in the late eighteenth and nineteenth centuries, and at the same time they provide social, historical, and moral insights. While instructional works of music were published in America as early as 1721, they continued to have a decidedly sacred flavor throughout the century. Secular collections were not abundant until the end of the eighteenth century. The earliest such work at the Society is probably Chauncey Langdon's *The Select Songster or a Collection of Elegant Songs With Music*, published in 1786, which fits the bill as both a songster and as secular music. The growing importance of secular music in the nineteenth century is manifested in the collection. As might be expected, both the volume and diversity of publications increased as the century progressed. By the mid-nineteenth century, works such as *Mendelssohn's Four-Part Songs* stand side by side with *Wood's Minstrel Songs* and the *Ole Bull Violin Instruction Book*.

Secular music for the 1821–90 period makes up the greatest number of works in the collection. Although not catalogued in the General Catalogue of the Society, all of these works are included in a checklist, entered by either author or title. Pre-1821 volumes of secular music are fully catalogued in the Imprints Catalogue.

The Society has an excellent collection of secondary sources concerning the history of music in America, including biographies, periodicals, reference works, and sound recordings. Two bibliographies are of particular importance. Oscar G. T. Sonneck's *A Bibliography of Early Secular American Music* (Washington, D.C. 1945), revised and enlarged by William T. Upton, is still the definitive work for secular music of the eighteenth century, providing locations of works and listings of the contents within collections. The Society's copy is annotated for recent acquisitions. Richard J. Wolfe's *Secular Music in America 1800–1825* (New York, 1964) is a vast, three-volume bibliography that also provides locations of works and is annotated with recent AAS holdings. [P.M.]

Songsters

According to the definition given by Irving Lowens in his *A Bibliography of Songsters Printed in America before 1821* (Worcester, 1976), a songster is a 'collection of three or more secular poems intended to be sung.' The Society's collection of songsters is one of the strongest in the nation. Of the more than 1,000 songsters housed at the Society, nearly 300 were published before 1821, while the remainder were published before 1877. The size of the collection is a result of continued acquisition over the years, including the purchase of the collection formed by H. Douglas Dana, a bookseller in Providence, Rhode Island, and the generosity of recent donors like Margery Morgan Lowens, who donated in 1983–84 the fine collections built by her late husband, Irving Lowens.

The songster was most often an inexpensive publication used as a popular source of poems, meant to be sung, of a romantic, patriotic, or comic nature; the songster was also used as a medium to express ever-changing topical concerns. A careful study of the materials would reveal interesting changes in the expressions of romantic love, patriotic fervor, and native humor during the years covered by the collection. Politics was a popular theme of many songsters, and a number were issued to advance the fortunes of specific political candidates. Social reform movements such as temperance utilized songsters to convey their message. Songsters compiled for the use of fraternal organizations such as the Freemasons and Odd Fellows are also commonplace in the collection. Another popular type contained songs about blacks. Printed predominantly in the northern cities during the 1840s and 1850s, they perpetuated racial stereotypes. The songster also provides an interesting resource for the study of popular poetry, as well as for the tunes to which the poems were meant to be sung. In many instances the name of the tune is listed with the poem.

The songsters printed before 1821 are shelved in the Dated Books and Dated Pamphlets collection and are fully catalogued in the Imprints Catalogue under the appropriate main entry, with additional access under title, subject heading, and the genre heading 'Songsters.' Additional entries are filed under the date and place of publication, and the name of the printer, publisher, or bookseller. Songsters published from 1821 to 1877 are shelved as one collection. Since the majority of songsters were compilations or the works of anonymous authors, the collection is accessed by title with added entries for authors. A checklist of these 1821–77 songsters was completed in 1980. The Lowens bibliography, cited above, is the standard reference work. [J.F.C.]

COOKBOOKS

Thanks to the generosity of Waldo Lincoln, who gave his personal library of American cookbooks to the Society in 1929, the researcher is blessed with a collection rich in depth and breadth. A 1985 estimate of its size sets the figure at around 1,100 volumes, for the most part published before the year 1877. The Society actively collects in this field whenever opportunities arise.

Although the Society's collecting policy emphasizes materials printed in the New World English colonies and the United States, a few cookbooks with British and other imprints will also be found here. This is not to be wondered at, since the first cookbook published in this country did not appear until 1742, when William Parks, a Williamsburg printer, gave the public *The Compleat Housewife*. This was a reprint of a London bestseller published fifteen years earlier. The Society's copy of this scarce book is in excellent condition. Another half-century was to elapse before the appearance of what is generally considered to be America's first truly indigenous cookbook, Amelia Simmons's *American Cookery*, published in Hartford in 1796. A copy of this work is also in the Society's library, as are several later editions, all rare.

Fifteen handwritten receipt books are located in the Manuscripts collection. They date from 1650 to the late nineteenth century. A few items germane to the general topic of domestic economy are shelved in the Graphic Arts Collection. In addition, the Society holds strong collections of printed books on the subject of domestic management theory, kitchens and their furnishings, the history of American eating habits, manuals of brewing, baking, and confectionery, not to mention biographies of such nineteenth-century giants as Catharine Beecher, Sarah Hale, and Lydia Child. These provide historical and sociological background that goes far beyond the realm of cookbooks narrowly defined.

Materials pertaining to the subject of cookery, including cookbooks, dated pamphlets, dated books, almanacs, and graphic arts, are contained in a comprehensive checklist made in 1983.

These are accessible by both author and title, and, in some cases, by subject. In addition, pre-1821 materials are to be found in the Imprints Catalogue. Researchers may also look in the General Catalogue for information under the subject heading 'Cookery' and related terms such as 'Food' and 'Diet.'

An additional help is Eleanor Lowenstein's *Bibliography of American Cookery Books 1742–1860*, published by the Society in 1972, and William Woys Weaver's 'Additions and Corrections to Lowenstein's *Bibliography of American Cookery Books, 1742–1860*,' which appeared in volume 93 of the *Proceedings* (1983).

Cookbooks refuse to remain in the kitchen, for when we look more closely, we find that they illuminate many other aspects of our past: technological (in the shift from fireplace to stove, from pump to running water), sociological (family composition and the relations between man and wife, parent and child, mistress of the house and servant), and religious and scientific (nutrition theories and fads). For students of women's history, cookbooks present a wealth of information about the changing role of women, without which our understanding of past and current ideals would be much impoverished. [A.C.M.]

TRADE CATALOGUES

Though seemingly ephemeral, trade catalogues are an invaluable primary source for the study of social, business, industrial, and advertising history. They have survived in abundance in many separate and, generally, uncatalogued collections around the country. The American Antiquarian Society's holdings include a small number of early trade catalogues through 1830 (price lists, circulars, special advertisements, for example), commensurate with the limited amount of trade material published during that era. The bookseller, the druggist, and the seedsman were the primary issuers of catalogues, and their contributions to America's trade history are fully catalogued in the Dated Books and Dated Pamphlets, Broadsides, and Reserve collections at AAS.

The bulk of the Society's Trade Catalogues collection, approx-

G. & D. COOK & Co's.
ILLUSTRATED CATALOGUE OF CARRIAGES
AND SPECIAL BUSINESS ADVERTISER
NEW HAVEN, Ct.
R. HOE & CO'S
LIST OF PRICES
EXTRA CAST STEEL SAWS,
HAMPSON WHITEHILL & CO.
STEAM ENGINES AND BOILERS
38 CORTLANDT ST. NEW YORK.
1875 Spring and Summer 1875
CATALOGUE
DEMOREST'S
PATTERNS
FASHIONS
GREAT REDUCTION IN PRICES
L. E. PERKINS'
51 MAIN STREET, TAUNTON.
STOCK MUST BE REDUCED!
E. BUTTERICK & CO'S PATTERNS, 555 BROADWAY, N. Y.
DOTY'S
Clothes Washer
THE PECULIARITIES OF THIS MACHINE:
1st. It is small. It occupies but little more room than a common Washtub.
2d. It is easily worked. A feeble woman may sit down and work it. A little girl can stand and operate the Washer.
3d. It secures all the well-known benefits of using suds boiling hot. The hand need not touch the water, either in washing or wringing, with the Universal Clothes Wringer attached; hence it cleanses ordinary linen in a few moments, at a single operation, and the worst parts of collars and wristbands so effectually, that seldom any, and never but little rubbing is necessary. There is no exaggeration in this statement, IF THE MACHINE IS USED ACCORDING TO DIRECTIONS.
4th. It has the important quality of doing its entire work without the slightest injury to the garment. It operates on the fulling-mill principle of pushing the mass together by gentle yet forcible pressure, the mass slightly rolling, and receiving each blow in a new place; all this, while immersed in the BOILING HOT suds, and the steam shut in.
DOTY'S WASHER
FAMILY SIZE.
Length, 28½ inches; Height, 28 inches.
Price, $14. "Extra" Wringer, $9.

imately 900 items, spans the years 1831 through 1876. Many of the catalogues are undated, and a few are twentieth-century reprints of earlier items. They run the gamut from tiny leaflets to bound folio volumes, and testimonials and references from previous owners are common. The collection also contains several lavishly illustrated items. The Society holds trade catalogues on such diverse subjects as farm implements, heavy industrial and engineering machinery, home amusement, furniture, sewing, and bridge construction. Even funeral casket devotees and soda fountain historians can find treasures here. The Society focuses on the acquisition of seed and nursery catalogues; now owning more than one hundred such items, with at least fifty firms represented.

Access to the Society's Trade Catalogues collection is through the General Catalogue and through a card-file checklist, arranged alphabetically by company name. This latter may be used in conjunction with Lawrence B. Romaine's subject-classified *Guide to American Trade Catalogues 1744–1900* (New York, 1960). Also available is a separate checklist of the Society's holdings of seedsmen's catalogues that was made in 1981 and is currently being updated.

Pertinent trade items held in the graphic arts department should also be mentioned. Two distinct collections to note are the William Allen Son's Co. scrapbook and the Centennial Collection of 1876. The William Allen Son's Co. scrapbook, given to the Society in 1981 by Mr. and Mrs. Arthur Knight, contains bills and circulars that deal chiefly with industrial machinery and heavy manufactures. A checklist of the contents of this collection is available for research use. In 1972, the Society purchased a collection of trade items obtained at the Centennial Exposition of 1876 from local bookdealer William Gavin. Although the bulk of this purchase is in the graphic arts department, a portion of it may be found in the Trade Catalogues collection proper. [P.A.B.]

NEWSPAPERS AND SERIALS

NEWSPAPERS

The American Antiquarian Society is this nation's chief repository for early American newspapers, and a significant portion of research done at the Society draws upon the Society's collection. The primary goal for the collection is to acquire, preserve, and make available for research newspapers published in the eighteenth and nineteenth centuries in the United States, Canada, and the English-speaking West Indies. To this end, the Society adds, through gift and purchase, an average of 3,000 issues a year to its holdings. Building on Isaiah Thomas's gift in 1812 of 382 titles in 551 volumes, the Society has accumulated over 15,000 newspaper titles in 20,000 volumes. Today, AAS has more than two million issues on five miles of shelving. The collection is preserved in protective folders and boxes in a climate-controlled environment. In 1973, to assure the preservation and usefulness of its newspaper holdings, the Society established the position of curator of newspapers and serials.

The collection is comprised of newspapers from all fifty states and the District of Columbia, the Canadian provinces, the West Indies, Great Britain, Mexico, and Latin America. British newspapers are maintained through the Revolutionary War period; newspapers from Latin America and the non-English-speaking Caribbean countries are retained but no longer acquired.

The Society's collection of pre-1821 American newspapers is the world's finest. It contains 1,494 titles or 71 percent of the 2,120 newspapers listed in Clarence S. Brigham's *History and Bibliography of American Newspapers, 1690–1820* (Worcester, 1947). The following newspapers are a sample of the Society's holdings for this period: *The Boston News-Letter*, the first continuously published newspaper on this continent; the *Pennyslvania Evening-Post*, printed in Philadelphia and the first attempt at an American daily newspaper; the *New-York Weekly Journal*, printed by Peter and Catherine Zenger from 1734 to 1751; and one of the two extant original copies of the oft-reprinted *Ulster County Gazette* from Kingston, New York, reporting on George Washington's death. AAS holds the only copy or copies of 118 of these early titles, among them *The Senator* (1814) from Washington, D.C., and the *Herald of Liberty* (1798–1802) from Washington, Pennsylvania. The post-1820 holdings are also among the best in the country. Although the general cutoff date for collecting at the Society is 1876, this date varies for states beyond the Mississippi, based on the commencement of printing in those areas. AAS collects newspapers from Arkansas, Kansas, Nebraska, Oklahoma, Oregon, and Texas to 1880; from Montana, North Dakota, South Dakota, Washington, and Wyoming to 1890; from Arizona, Colorado, Idaho, New Mexico, Nevada, and Utah to 1895; and from Alaska to 1900.

Not surprisingly, the depth and breadth of newspaper holdings at AAS are greatest for the eastern United States. These include 1,588 titles for Massachusetts, 1,323 for Pennsylvania, 1,493 for New York, 844 for Ohio, 300 or more from Georgia, California, Illinois, Indiana, Maine, Michigan, Missouri, New Hampshire, Tennessee, and Wisconsin, over 200 from Georgia, Connecticut, Kentucky, Louisiana, Maryland, Mississippi, New Jersey, Rhode Island, and Vermont, and over 100 each for Colorado, Minnesota, Nebraska, North Carolina, South Carolina, Texas, West Virginia, and the District of Columbia. While the Society has many titles for a few states west of the Mississippi such as Iowa (446) and Kansas (176), the runs for those titles are limited to one or two issues.

The *Newspaper Cataloging Manual* of the Library of Congress defines a newspaper as a serial publication designed to be a primary source of written information on current events connected with public affairs, either local, national, or international, not limited to a specific subject matter. The Society, however, collects every kind of newspaper, those that fit the definition strictly, those that are really periodicals in newspaper format, such as religious, temperance, college, or literary newspapers, and those that do not seem to fit either category, including campaign, advertising, church fair, or price-current newspapers. Geographically, the collection encompasses most areas on the North American continent and the contiguous islands.

The collection includes many ethnic newspapers. Of the 5,000 newspapers and periodicals listed in Karl Arndt and May Olson's *The German-Language Press in the Americas, 1732–1955* (Munich, 1980), the Society has 451 of the pre-1877 titles. Its holdings also contain 71 of the 169 titles in Edward Laroque Tinker's *Bibliography of French Newspapers and Periodicals in Louisiana* (Worcester, 1933) and 20 of the 35 pre-1877 titles in Brown's *Checklist of Negro Newspapers in the United States* (Jefferson City, Mo., 1946).

Wallpaper newspapers form an unusual group in the Society's holdings. These were newspapers printed on the back of wallpaper samples because of the paper shortage in the southern states during the Civil War. The Society has three copies, each on different wallpaper, of the most famous of these newspapers, the *Daily Citizen* from Vicksburg, Mississippi, for July 2 and 4, 1863. It holds twenty-two of the thirty-two titles listed by Clarence Brigham in his essay on the subject in *Bibliographical Essays: A Tribute to Wilberforce Eames* (Cambridge, Mass., 1924).

The newspaper collection of the Society has grown and continues to grow through purchases, particularly of pre-1821 issues and of issues from the western states for which its holdings are weak, as well as through gifts from individuals and from institutions. Many of the Louisiana newspapers came from Edward Laroque Tinker, while Waldo Lincoln brought together the West Indian issues. The Society's Rowell collection was assembled over a number of years. G. P. Rowell, noted for his annual newspaper directories, organized an exhibition of all current newspapers in the United States for the Philadelphia Centennial Exposition in 1876. Discards from the exhibition found their way into the collections of the Chicago Historical Society. In 1915, that society donated the issues from the eastern states to AAS, and in 1974 it completed the gift with all its Rowell newspapers except for those from Illinois and Indiana. After World War II, New

England institutions sent hundreds of newspapers to the Society. In 1973, the Society circulated to historical societies, public libraries, and academic institutions a request for those newspapers they could not maintain.

Providing bibliographical control of and access to its research materials has been a major activity of the Society throughout its history. This has been as true for newspapers as for any other group of materials at the Society, from Thomas's catalogue of his gift to the recent participation by AAS in the United States Newspaper Program, underwritten by the National Endowment for the Humanities.

Because the collections of the Society are shelved in areas closed to the public, researchers gain access to them through the General Catalogue, bibliographical tools, and terminals linked to national databases. The catalogue provides holdings information rather than bibliographical descriptions. It is divided into pre-1821 and post-1820 sections, with each section arranged alphabetically by state, town, and title. The first section contains issue-by-issue records of the Society's early holdings. A major project in process is the collating of the post-1820 newspapers to provide similar information for the second section of the catalogue. Because of new acquisitions and cataloguing, the General Catalogue will remain the most up-to-date guide to the Society's holdings.

As a participant in the United States Newspaper Program, the Society entered bibliographical and holdings records for 14,000 of its pre-1877 American titles into the national database, OCLC. With the completion of this project, AAS will continue to have access to these records through RLIN.

The Society's reference and secondary collections contain the works necessary to support research in its newspaper holdings. These holdings are found in Brigham, Winifred Gregory's *American Newspapers, 1821–1936* (New York, 1937), *Newspapers in Microform, 1964–1985* (Washington, D.C., 1985), *The USNP National Union List* (Dublin, Ohio, 1985), Waldo Lincoln's *Bibliography of West Indian Newspapers in the American Antiquarian Society* (Worcester, Mass., 1926), and Steven M. Charno's *Latin-American Newspapers in American Libraries* (Austin, Tex., 1968), as well as the 'Checklist of Campaign Newspapers in the American Antiquarian Society.' State, county, and town histories found in the Society's local history collection provide descriptions of newspapers, as do histories of journalism and of specific newspapers.

The lack of indexes to early American newspapers often hampers research. Brigham's *Bibliography* contains a title index and is complemented by Edward Connery Lathem's *Chronological Tables of American Newspapers, 1690–1820* (Worcester, 1972). Avis G. Clarke's typescript 'An Alphabetical Index to the Titles in American Newspapers, 1821–1936' is a useful tool for nineteenth-century research. The few subject indexes to eighteenth- and nineteenth-century newspapers include Lester Cappon's *Virginia Gazette Index* (Williamsburg, Va., 1930) and the WPA index to the *Hampshire Gazette* of Northampton, Massachusetts (Boston, Mass., 1939), as well as more well-known indexes held by AAS. In general, newspaper indexes are genealogical in nature, such as the *Index of Obituaries in Boston Newspapers, 1704–1800*, compiled by the Boston Athenaeum (Boston, 1968) and the 'Index to Marriages and Deaths in *Columbian Centinel*,' a typescript at the Society. Indexes in the Society's own collections are listed in the 'Checklist of Newspaper Indexes in the American Antiquarian Society.'

Because the Society's primary interest is in original materials, it seldom purchases newspapers on microfilm, and films titles from its own collection only on request from a patron. As part of its cooperation with Readex Microprint Corporation, however, the Society has acquired a complete set of the Readex Early American Newspaper Series on microcards and microfilm. The small collection of microfilm accumulated through the years includes two recent additions of Worcester titles, *The Massachusetts Spy*, 1821–1904, and *The Aegis*, 1801–97. To preserve originals, AAS has also acquired facsimile reprints of titles such as *The Boston News-Letter*, *The Newport Mercury*, and *The New-York Gazette*, among others.

Although the newspaper collection of the Society is heavily

used, historians and researchers have yet to realize its full potential. As Clarence Brigham noted in his *History and Bibliography of American Newspapers*, 'If all the printed sources of history for a certain century or decade had to be destroyed to save one, that which could be chosen with the greatest value to posterity would be a file of an important newspaper.' [J.A.T.]

THE PERIODICAL COLLECTION

With an outstanding collection of early American periodicals, the Society offers researchers many opportunities for studying the thought, culture, and life of North America through contemporary eyes. As with many of the Society's collections, Isaiah Thomas's personal library formed the basis of the Society's holdings of periodicals, now numbering around 6,000 titles in 55,000 volumes. A scholar can find periodicals published in the United States, Canada, Great Britain, or in Turkey by American missionaries. Although the holdings are generally limited to titles published before 1877, the cutoff date is extended, as it is in the newspaper collection for those parts of the United States in which printing commenced at a later period. The Society continues to acquire periodicals published in this period through purchase and gifts. It also subscribes to about 650 current periodicals issued by state, county, and local historical associations, as well as by institutions and publishers on American history, culture, and the arts.

Nearly all the eighteenth-century American and Canadian periodicals are represented, as well as a very large percentage of those titles issued before 1820; also available are extensive files of ephemeral and important journals from 1821 to 1876. Unusual and short-lived magazines can be found in the Society's collection along with better-known titles with long runs: one of the first American periodicals, Benjamin Franklin's 1741 *General Magazine*; the first Massachusetts periodical, *The Boston Weekly Magazine* of 1743; or the first American children's periodical, *The Children's*

Magazine of 1789. Among the ethnic publications that are found in the Society's collection are the Sioux missionary publication, *Iapi Oaye*, of the 1870s; *L'Album Litteraire*, issued in New Orleans in 1843 by young French-speaking black men; and the Welsh *Y Cyfaill* published in Utica, New York, dating from 1843.

The collection covers a multitude of subjects: anthropology, antislavery, archaeology, education, fashion, literature, medicine, music, photography, printing, prison reform, religion, science, sport, and temperance. The best run of *The Home Journal*, the predecessor to the present day *Town and Country*, is on the Society's shelves, as well as issues of *The New Orleans Medical and Surgical Journal* (1845–76), the feminist *Revolution* (1868–72), and the notorious *Woodhull and Claflin's Weekly* (1870–76). The holdings contain the first English-language periodical in Canada, *The Nova Scotia Magazine* (1791–92) and British magazines such as *Blackwood's Edinburgh Magazine* (1817–1904). A special group of religious periodicals are those published by the Adventists, considered one of the better collections outside Adventist institutions.

Information about periodical holdings at AAS is found in the Society's copy of the *Union List of Serials*, annotated with up-to-date acquisitions data. There is at present no card catalogue for periodicals, although titles shelved with the collections of learned societies, local history, and institutions have cards in the General Catalogue. The holdings of current journals can also be found in the visible index, a daily record of incoming issues. AAS holdings for current subscriptions have been entered into OCLC and RLIN. The Society keeps RLIN up-to-date and is cataloguing its early titles online. The *Worcester Area Cooperating Libraries* (WACL) *Union List* is available online through the OCLC database and in printed format. *The Directory of State and Local History Periodicals* (Chicago, 1977) is also annotated with each new added title.

Another means of locating periodicals in the Society's collection is through in-house checklists such as the 'Checklist of American Temperance Periodicals' and the 'Checklist of American Children's Periodicals.' Some published subject bibliographies

ANTIQUES
The Journal of Arizona History
The Atlanta Historical Bulletin
The California
HORTICULTURIST
Floral Magazine.
NOVEMBER, 1870.

are also annotated with AAS holdings. Among these are James Danky's *Women's Periodicals and Newspapers from the Eighteenth Century to 1981* (Boston, 1982) and his *Native American Periodicals and Newspapers, 1828–1952* (Westport, Conn., 1984). All important subject bibliographies are in the Society's reference collection including Arndt and Olson's *The German Language Press in the Americas* (Munich, 1980), and Eugene Willging and Herta Hartfeld's *Catholic Serials of the Nineteenth Century in the United States* (Georgetown, D.C. 1968).

Although the bulk of the Society's periodicals is shelved in one part of the stacks, periodicals for local history, learned societies, and institutions are shelved with other materials of the same type. Those in newspaper format are placed with newspapers. An inventory of the collection is being constructed that will provide a card catalogue that indicates the location of each title and provides detailed holdings information.

The Society acquires all periodical indexes relevant to its collection. One of the major indexes is the *Early American Periodical Index, 1743–1850*, compiled by the WPA and now available on Readex Microprint cards. It is comprised of several indexes for authors, subjects, titles, poetry, and book reviews in articles found in 370 titles. The microfilm collection of periodicals is small but does include the American Periodical Series for 1741–1835 (Ann Arbor, 1979). AAS microform holdings are listed in a card file, with master negatives also entered into RLIN.

[J.A.T. and A.T.Z.]

AMATEUR NEWSPAPERS

The amateur newspaper occupies an unusual place in the history of journalism. An amateur journal is a periodical created to afford pleasure to its readers and to its publisher. The rage to publish, rather than profit, is the motive that most often induces people to become journalists; and throughout its history, most but not all amateur journalists have been juveniles.

THE FLOWER
The Casket
THE BOSTONIAN
THE YOUNG AMERICAN
THE JUG
MONTHLY NEWS
Salutatory
THE THESPIAN MIRROR
To the Public
EXTRA
OUR HOPE
THE BEE
THE MONUMENT
BANNER.
HARPSWELL, MAINE, APRIL 21 1860
The Advertiser.
THE ADVENTURES OF DICK ALSTON.
The Centennial
PATRONS OF THE

The amateur newspaper collection at the American Antiquarian Society consists of about 50,000 issues. There are more than 5,500 titles from every state except Alaska and Hawaii, as well as issues from fifteen foreign countries, thus making the Society's holdings among the largest and most extensive in the United States. The Society has files of seventy-eight titles from Worcester alone.

Amateur newspapers range in size from miniature to quarto, and from two to as many as thirty-two pages. The Society concentrates its efforts on acquiring amateur papers that fall within the same time period as that for the regular newspaper collection.

It is not known for certain when amateur journalism began, but the first amateur newspaper published in the United States is believed to be the *Thespian Mirror.* This paper was edited by John Howard Payne, the future dramatist and actor, while he was a New York City bookkeeper. The first issue appeared on December 28, 1805, and the final issue on May 31, 1806. The journal was printed for Payne, then just fourteen years old, by professional printers and was intended for an adult audience. AAS owns seven of the fourteen issues of this amateur paper that included theatrical reviews, biographical sketches, and poetry. Pen-printed amateur newspapers were also made by amateur journalists. The Society holds the earliest extant issue of a pen-printed amateur newspaper by James Johns of Huntington, Vermont, dated October 10, 1834. During the 1840s and 1850s, amateur newspapers began to proliferate. Until the late 1860s there were three methods of publication available to the amateur journalists: (1) writing or printing the contents of the paper with pen or pencil, the method used by the editors of the *Flower,* published at Smithfield, Rhode Island, in 1836, and for the *Casket* , published at Boylston, Massachusetts, in 1857; (2) paying a professional printer to do the typesetting and presswork with the aid of the amateur's manuscript, as did John Howard Payne; and (3) building his or her own press, as did Marcus Rogers, who in 1854 constructed a press and printed the *Rising Sun* at Mill River, Massachusetts, and Cyrus Curtis, who in 1865 utilized a discarded hand press to print the *Young American* at Portland, Maine.

In 1867, Benjamin O. Woods of Boston invented an inexpensive 'Novelty Press' (as Woods named it), which, due to its simple construction, could be sold for a few dollars. Woods advertised his press in periodicals designed for youngsters, which also carried news about amateur printers and editors.

During the ten years following the invention of Wood's press, the number of amateur newspapers in existence increased from fewer than 100 to almost 1,000. Although many of these were short-lived, some endured for five years or more. Amateur journalism flourished all across the country from Maine to California. In addition to editorials, original short fiction, essays, jokes, stories gleaned from other publications (both amateur and professional), and poetry, puzzle departments began to appear as a regular feature in many amateur papers. The Society has a great many amateur journals from the golden age of the 1870s and 1880s.

The Society's amateur newspaper collection is arranged in 134 folio boxes. Four boxes shelved at the beginning of the collection contain general information about amateur journalism. In the next 128 boxes are the amateur newspapers filed alphabetically by title. The last two boxes hold uncatalogued brochures and books that were written and/or published by amateurs.

Five trays of catalogue cards provide access to the amateur newspaper collection. Each card lists the place of publication, the title of the newspaper, its frequency of publication, and the Society's holdings of the title. The catalogue cards are filed alphabetically by place of publication. After the United States titles, which take up four and a half trays, there are about fifty cards that list titles with unidentified places of publication; these are filed alphabetically by title. The final cards list the foreign amateur newspapers, which are filed by name of country, city or town, and alphabetically by title. [D.R.L.]

MANUSCRIPTS AND ARCHIVES

MANUSCRIPTS

AAS manuscript collections are rich resources for the study of American history and culture. Numbering over 1,200 collections, the Society's manuscripts span the years 1613–1930 and are useful to scholars working in a variety of disciplines. An excellent account of the Society's acquisition of manuscripts, by William L. Joyce, appears in the Society's *Proceedings* 89 (1979):123–52. The history of AAS's manuscript collecting is varied, although the underlying principle for acquisitions has always been that material should relate to the history of America. Over time, the acquisition of manuscripts has become increasingly focused. As the Society has moved away from the role of general library and museum to that of research library, so too has the gathering of manuscripts become more specialized. Today, manuscripts are actively acquired in four areas of collection strength: American book publishing and collecting; New England diaries; papers of prominent early New Englanders in the political, religious, and military spheres; and selected papers and records of eighteenth- and nineteenth-century Central Massachusetts families and voluntary associations. Manuscript collections, like their printed counterparts, are generally limited chronologically to the period before 1877.

The early American book trades are a central focus of the entire Society. Isaiah Thomas's gift of his own papers initiated the Society's acquisition of book trade manuscripts. Through the years, large and small collections of records of publishing businesses have been acquired, including those of Mathew Carey, Copeland & Day, D. C. Heath, G. & C. Merriam Company, Lee & Shepard, McCarty and Davis, and West, Richardson, & Lord. Booksellers' records, such as those of William Cobbett, Jeremiah Condy, and the Boston Booksellers Association, were also acquired, as were records of bookbinders like William Merriam and printing press manufacturers such as R. Hoe and Co.

The Society's diary holdings are made particularly strong by several extensive runs of diaries. Those of the Massachusetts silhouette artist Ruth Henshaw Bascom, for example, number fifty-two volumes and span the years 1789–1848; the diaries of William Bentley, the Salem, Massachusetts, minister, cover the years 1785–1819; Boston minister Andrew Bigelow's diaries number sixty-six volumes and span the years 1816–76; those of Susan E. Forbes, a Massachusetts teacher and housewife, cover the years 1841–1908 in sixty-six volumes; Westborough, Massachusetts, minister Ebenezer Parkman's diaries span fifty-five years, 1723–1778; and those of Caroline Barrett White, a teacher and housewife from Massachusetts, number thirty volumes and cover the years 1849–1915.

Virtually all phases of Massachusetts history are covered by the Society's manuscript collections, including personal and family, religious, business, political, social, early industrial, and military themes. From the outset, the Society was concerned with every aspect of American history, and archaeology was a major interest of early members. Although the artifacts that accumulated during the nineteenth century have long since been transferred to other institutions, several manuscript collections (and the Society's own archival records) reflect this interest in archaeology. The Society also has several collections of manuscript music, strengthened by a recent accession from Mrs. Irving Lowens.

Family papers constitute a significant component of the AAS manuscript collection. A variety of collections, primarily of Central Massachusetts families, document eighteenth- and nineteenth-century life in New England. Included among the Society's substantial holdings of family papers are those of a free-Black family in Worcester named Brown, the Chandler-Ward families of Petersham and Lancaster, the Lincolns of Worcester, the Parkman family of Westborough, three generations of Worcester Salisburys, the Wards of Shrewsbury, and the Waters family of Millbury.

The Society's collection of voluntary association records complements its other manuscript holdings. Charitable organizations, literary societies, and musical associations were popular eighteenth- and nineteenth-century commitments and diversions. The Society's records of such organizations add to the documentation of that period of New England culture.

The Society's manuscript collections were arranged and described under the supervision of William L. Joyce during the 1970s with funding from the National Endowment for the Humanities. The resulting intellectual control of the collections consists of two parts, the card catalogue and collection descriptions. The card catalogue was published in four volumes in 1979 by G. K. Hall as *Catalogue of the Manuscript Collections of the American Antiquarian Society.* It is available for use in many university and research libraries in the United States, Canada, and Great Britain. The *Catalogue* is arranged alphabetically, with personal names, geographic names, and topical subject entries interfiled in one sequence. At the beginning of volume one is a list of the subject entries used when processing the collections and a list of those collections that were item-catalogued (meaning that each manuscript in the collection was described individually). Typically, the following information is provided for each item: collection name, author name, recipient name, date, type of manuscript, place where the manuscript was written, number of pages, and a very brief summary of content. Most of volume four is a chronologically arranged list of materials in the item-catalogued collections.

All collections are represented in the *Catalogue* by a Collection Description card, so marked in the upper right-hand corner. This card is followed by a second card, on which is listed all subject headings used in cataloguing that collection. Each of these subject headings is also typed in capital letters at the top of copies of the Collection Description card. These subject headings are filed in their appropriate place in alphabetical sequence. Item-catalogued collections are so noted in the left-hand margin of the Collection Description card. The collection name of item-catalogued collections appears in the upper left-hand corner of all cards.

The *Catalogue* also serves as an index to the Society's second level of manuscript description, collection descriptions. Filed alphabetically in binders at the Society, collection descriptions are available for all processed collections and contain the following information: collection name, size, location, *National Union Catalog of Manuscript Collections* number (when applicable), finding aids, source, and biographical information and content description. If a collection is available on microfilm, that information is stamped on the collection description. Large collections have longer collection descriptions, which often include contents lists. Contents lists itemize collections at the box, folder, or volume level, providing more specific access to the information contained within collections.

A third means of access to some AAS manuscripts is through the RLIN database. Since mid-1985, RLIN Archival and Manuscripts Control (AMC) records have been created for manuscript accessions and, retrospectively, for some collections. As new collections are processed, AMC records for them are created. The amount of information about a collection contained within an AMC record can vary greatly, but the minimum bibliographical information contained in all AMC records is the name of the individual or entity that created the material, title and dates of the collection (for example, 'Papers, 1823–1856'), and physical extent of the collection (for example, '3 boxes'). AMC records for AAS collections that were acquired or revised since publication of the *Catalogue* are gradually being added to the RLIN database with the goal of having all AAS manuscript collections described in either the *Catalogue* or RLIN.

The Society's Miscellaneous Manuscripts collection is uncatalogued. It consists, for the most part, of single items by a wide variety of individuals, businesses, and town governments. Occasionally, there are several items by one person, not enough to warrant making a separate collection. Arranged alphabetically by personal name or place name, with a few subjects included, the Miscellaneous Manuscripts collection has been catalogued through the letter A. An alphabetical checklist has been compiled to provide some access to the entire collection. [B.T.S.]

ARCHIVES

The Society's archival records, 1812-present, are currently being arranged and described for research use. At the end of the project, a finding aid will be generated to facilitate use of these records. Included in the finding aid will be an administrative history of the Society, a description of the types of information found in the records, a list of folder and volume headings, and correspondent and subject indexes.

In addition to documenting the workings of the Society itself, AAS records are useful to scholars of nineteenth- and twentieth-century American life, particularly those examining such topics as the rise of learned institutions, the growth of research collections and patterns of collecting, bibliography, philanthropy, ethnology, archaeology, specific Society members, and United States, Massachusetts, and Worcester history.

Since its founding, the Society has actively contributed to the understanding of this country's history by its commitment to accumulating, preserving, and disseminating information about the United States, and by aiding scholarly research in its collections. The Society's history mirrors that of the United States; AAS grew and became a more professional organization, reflecting the intellectual changes taking place in this country. A broad spectrum of persons have been members of AAS, and their participation in the Society, or lack thereof, has influenced the direction and growth of the Society. Indeed, some members have been particularly influential in the development and growth of this country.

The records of the Society are organized in series according to function, and within series they are arranged in chronological order. Correspondence is arranged chronologically by decade, and, within each decade, is arranged alphabetically by correspondent. In addition to correspondence, the AAS archives contain meeting minutes of the Council of the Society, accession records, financial records, library catalogues, publications, and visitors logs. The Society's records through 1939 are open for research use. For use of post-1939 records, researchers should consult with the curator of manuscripts. [B.T.S.]

Programs of the Society

DEPARTMENT OF RESEARCH AND PUBLICATION

The Department of Research and Publication oversees the Society's publishing program, its organized research activities, its program of fellowships for visiting scholars, and several educational activities aimed at constituencies of professional scholars, undergraduate and graduate students, and the general public. These myriad activities depend for their existence on the library collections of the Society. In essence, they fulfill the third major element in the Society's founding mandate—that is, to make the collections available for use and to disseminate the fruits of research carried on in the Society's library. Some of the activities that the department supervises, such as publishing, are ancient ones for the Society, while others, such as formal educational activities, are more recent developments. [J.B.H.]

PUBLICATIONS

Isaiah Thomas, quite appropriately, established a touchstone for all of the publications of the American Antiquarian Society with his *The History of Printing in America*, first published in 1810. He defined both the value of and the audience for such an endeavor in his introductory remark that 'the art which is the preserver of all arts, is worthy of the attention of the learned and the curious.' Indeed, it was Thomas's conviction that detailing the history of printing in America to the time of the Revolution, a field in which he himself had half a century's experience to draw upon, would serve an enduring national purpose. He stated in the preface to

the *History of Printing*, that his chronicle could serve to 'convey to posterity a correct account of the manner in which we have grown to be an independent people, and can delineate the progress of the useful and polite arts among us with a degree of certainty which cannot be attained by the nations of the old world.' This conviction accorded well with Thomas's larger patriotic endeavor, the founding of the American Antiquarian Society as a library that would both preserve America's heritage and provide new perspectives on that national history. Through its publications, the Society continues to fulfill that ideal.

Among the earliest publications by the Society were the records of the annual meetings. The first 'Account of the American Antiquarian Society' was printed in 1813. It included a petition to the Massachusetts legislature to establish the Society, the act of incorporation, the laws of the Society, and the record of that first meeting. Subsequent sporadic publication of the records of meetings essentially consisted of votes passed and elections of officers and members; occasionally, activities of the meetings appeared in newspapers rather than as Society-sponsored publications. In 1843, there was a serious effort to issue the *Proceedings of the American Antiquarian Society* in a regular series, and two numbers were printed. However, consecutive publication of the formally titled *Proceedings of the American Antiquarian Society* was not firmly established until 1849.

The Society did not ignore scholarly research, however. The efforts of the committee on publication in 1819 led to the establishment of the series *The Transactions and Collections of the American Antiquarian Society*. This series bore the subtitle *Archaeologia Americana*, a reflection of the journal published by the Society of Antiquaries of London, the organization that had been the model for the founders of the American Antiquarian Society. Volume I, printed in 1820, contained accounts of the discovery of the Mississippi and 'Conjectures Respecting the Ancient Inhabitants of North America.' Volume II appeared in 1836 and presented studies of the Indian tribes in North America by Albert Gallatin and Daniel Gookin. Volume III concerned itself with colonial

American history and included the Records of the Colony of Massachusetts Bay and the diaries of John Hull, Thomas Winthrop, and Thomas Lechford. Later volumes reprinted the diaries of Christopher Columbus Baldwin and Isaiah Thomas. This series also reprinted Thomas's *History of Printing* in two volumes in 1874. Twelve volumes appeared intermittently until 1911.

The *Proceedings of the American Antiquarian Society* continued to appear alongside the *Transactions*, although much more regularly. In part, increasing publishing costs spelled the demise of the *Transactions* after 1911. After 1849 the *Proceedings* were published twice each year, keyed to the October and April meetings of the Society, and collected as an annual volume. In 1880, a 'new series' of the *Proceedings* began. With this series the *Proceedings* became primarily a learned journal, a change derived in part from the nature of the lectures delivered at the annual meetings of the Society.

Today, the *Proceedings* continues to fulfill the founders' goal by publishing the fruits of scholarly research. The journal is now less an account of the meetings of the Society, although these transactions are published in each issue, than a continuing record of the scholarly endeavors of the Society's members, staff, fellows, and other researchers. The *Proceedings* accepts articles in the general field of American history and culture through the year 1876, with particular emphasis on bibliographies, primary sources, and basic research tools. The Society routinely issues *Proceedings* articles as separate offprints. Offprints are available either directly from the Society or from the Society's book distributor, the University Press of Virginia.

Books published by the Society tend to deal with the printed record of the United States, and include bibliographies, source documents, and other materials that serve as tools for researchers. A number of these books have become the standard bibliographies in their fields. The Society has published the two-volume *History and Bibliography of American Newspapers, 1690–1820*, by Clarence S. Brigham (1947), volume 13 of Evans's *American Bibliography*, by Clifford K. Shipton (1955), *Vermont Imprints, 1778–1820*, by Marcus A. McCorison (1963), *A Descriptive Checklist of Book Catalogues Separately Printed in America, 1693–1800*, by Robert B. Winans (1981), *A Bibliography of American Children's Books Printed Prior to 1821*, by d'Alté A. Welch (1972), *A Bibliography of American Cookery Books, 1742–1860*, by Eleanor Lowenstein (1972), and *A Dictionary of Colonial American Printers' Ornaments and Illustrations*, by Elizabeth Carroll Reilly (1975). Two recent publications also highlight the Society's collections of early American bookbindings: *Bookbinding in Early America: Seven Essays on Masters and Methods*, by Hannah D. French (1986) and the revised edition of

the 1972 exhibition catalogue *Early American Bookbindings from the Collection of Michael Papantonio* (1986). In addition, the Society has published collections of monographs in areas relating to the history of the book in American culture. Among these are *The Press and the American Revolution* (1980), *Printing and Society in Early America* (1983), and *Needs and Opportunities in the History of the Book: America, 1639–1876* (1987).

The Society participated with the Readex Microprint Corporation on a long-term major project, entitled *Early American Imprints*, to publish in microform the nonserial material issued in this country from 1639 to 1820—some ninety thousand books, pamphlets, almanacs, and broadsides. Material in the first series of this project, covering the years 1639 through 1800, was edited by Clifford K. Shipton and is based on Charles Evans's *American Bibliography* (1903–55) and Roger Bristol's *Supplement* (1970). This is now available in Microprint and microfiche formats. A by-product of the publication of this series was the *National Index of American Imprints Through 1800: The Short Title Evans*, published by AAS in 1969. Material in the second series of this project, covering the years 1801 through 1819 and based on Shaw and Shoemaker's *American Bibliography* (1958–63), was edited by Shipton, James E. Mooney, and the present associate director for research and publication. Sets of this massive research resource exist in libraries both in the United States and throughout the world.

While the Society stocks its publications for sale at the library, most orders for books and *Proceedings* offprints are handled through the Society's book distributor, the University Press of Virginia, Box 3608, University Station, Charlottesville, Virginia 22903. Subscriptions to the *Proceedings* may be entered directly with the Society; libraries may subscribe through their usual subscription agencies. Back issues and reprints of the *Proceedings* are available from the Kraus Reprint Company, Route 100, Millwood, New York 10546. Recent back issues of the *Proceedings* may be ordered from the Society. A list of the Society's publications in print is available upon request. [M.S.McA.]

FELLOWSHIPS

The Society's Council in 1970 authorized the establishment at AAS of a program of funded visiting research fellowships in order to make more readily available its unparalleled resources in American history and culture. In the intervening years, the Society has awarded support to nearly two hundred scholars from most of the fifty states and from five foreign countries, including the People's Republic of China. Seventeen or eighteen fellows are now appointed annually. Many of the fellows have produced significant publications based in part on their mining of the rich resources of the AAS collections. Visiting scholars range from promising young graduate students embarked on their doctoral dissertations to some of the most accomplished writers of American historical literature. Their work has enriched humanistic research and teaching in America as well as a whole generation of AAS staff members. Perhaps no other program at AAS has done more to increase the American research community's knowledge of what is available for historical research at the Society. It has certainly brought within the orbit of the Society a group of very energetic, talented, and promising people. In a considerable number of cases, the fellowship at AAS has been but a prelude to ever greater involvement in the Society's affairs. In a very real sense, the former fellows have become the Society's alumni.

The first AAS fellowships were awarded in 1972–73. They were short-term grants, with modest stipends, that helped defray the expense of scholars in spending from one to three months in residence at the Society. In 1976–77, funds became available to provide fellowships at income-replacement levels to allow researchers access to AAS collections for periods ranging from six months to a full year.

The general short-term fellowships have been supported by a variety of funding sources and have been known by several names. Now they carry the name Kate B. and Hall J. Peterson Fellowships, and are funded by income on an endowment provided by Mr. Peterson, an AAS member, and his wife. Early in this decade the Society added two specialized short-term fellowship categories. The Albert Boni Fellowship for research in the field of early American bibliography and printing and publishing history is made possible by gifts from members of the family of the late Mr. Boni, a distinguished literary publisher and founder of Readex Microprint Corporation. The Frances Hiatt Fellowships for graduate students at work on their doctoral dissertations derive from annual gifts made by Jacob Hiatt, a Worcester industrialist and AAS member, in honor of his late wife. A grant from the Northeast Modern Language Association will permit the inauguration of a fellowship for literary studies at AAS in 1988–89. The AAS long-term fellowships have been made possible by a sequence of seven grants from the National Endowment for the Humanities. AAS is one of some sixteen independent 'centers for advanced study' funded under this NEH program. From these grants, the Society has appointed annually from one to four American Antiquarian Society-National Endowment for the Humanities Fellows. The Society also appoints Research Associates, scholars with third-party support, who receive no stipend from AAS but receive fellowship privileges.

Among the important books that have stemmed from AAS fellowships are *The Transformation of Political Culture: Massachusetts Parties 1790s–1840s*, by Ronald P. Formisano (New York, 1983); *Disorderly Conduct: Visions of Gender in Victorian America*, by Carroll Smith-Rosenberg (New York, 1985); *A Calculating People: The Spread of Numeracy in Early America*, by Patricia Cline Cohen (Chicago, 1982); *Soldiers of Light and Love: Northern Teachers and Georgia Blacks 1865–72*, by Jacqueline Jones (Chapel Hill, 1980); and *The Coming of Industrial Order: Town and Factory Life in Rural Massachusetts, 1810–1860*, by Jonathan D. Prude (New York, 1980).

The opening to fellows and other visiting researchers of the Goddard-Daniels House in the spring of 1982 has greatly enhanced the collegial aspect of the Society's fellowship program.

[J.B.H.]

EDUCATION

In the mid-1970s, the Society's Council appointed an ad hoc planning committee to study ways in which the Society could become more useful to various constituencies. It recommended, among other things, the establishment of formal educational activities for professional scholars, graduate students, Worcester area undergraduates, and the general, lay public. A five-year grant from The Andrew W. Mellon Foundation provided the funds for a part-time education officer and seed money to experiment with a variety of educational initiatives.

The first major undertaking was the establishment of an American Studies seminar for a select group of undergraduates from the five four-year colleges and universities in Worcester: Assumption, Clark, Holy Cross, Worcester Polytechnic Institute, and Worcester State. A committee made up of representatives from each of the participating institutions was formed to advise the AAS staff and to select the students who matriculate in the seminar. Thus, the Society is able to make the library collections available on a controlled basis to college undergraduates, a constituency that the Society ordinarily does not serve. The first seminar met in the fall of 1978. Stephen Nissenbaum, a University of Massachusetts historian, led that inaugural seminar whose theme was 'Writers Confront the Marketplace: Literature and Society in Jacksonian America.' A renewal of the seminar has taken place every fall since, each led by a different teacher in charge. Topics under scrutiny have ranged from popular culture in colonial America to the myth of violence in the late-nineteenth-century American West. The seminars have been successful educational activities for the students involved, have enriched the curricular offerings of the participating colleges, and have been both a challenge and great fun to the members of the library staff, who relish their role in initiating talented and eager students into the mysteries of research in a major library.

A number of activities provide opportunities for scholarly discussion and collegiality among faculty members and advanced graduate students within the region surrounding AAS. The earliest of these predated the actual establishment of the Society's education office. Two historians from nearby institutions, David Hackett Fischer of Brandeis and Ronald P. Formisano of Clark, asked AAS to serve as host and meeting place for periodical gatherings of scholars in the region active in research areas akin to their own. As a result, the Seminar in American Political and Social History has met at AAS some five or six times a year ever since. Seminars are held at the Goddard-Daniels House and followed by catered dinners (at moderate cost) in its elegant dining room. Several years ago, AAS established a similar Seminar in American Literary History, which draws participants from much the same geographical region. Over the years, distinguished scholars have given papers, mostly describing work in progress, at one or another of these seminar gatherings. They include Gordon Wood, J. R. Pole, John Murrin, Stephen Botein, Karen Kupperman, Alden Vaughan, Mary Beth Norton, Sacvan Bercovitch, and Leo Marx. Members of the political and social history seminar heard a preview of the Bancroft Prize-winning book *The Minutemen and Their World* from then-Ph.D. candidate Robert A. Gross. These seminar series all in all have allowed AAS to serve as a gathering place for scholars in the region and has helped to introduce the Society's collections and programs to them.

In addition to the seminar series, other AAS activities provide opportunities for professional scholars and graduate students in the area. These include informal brown-bag, lunchtime colloquia as well as occasional evening lectures on scholarly topics.

Within the last decade, AAS has also devised educational activities to serve the general lay public in the greater Worcester area. These have more often than not taken the form of a series of public lectures on a given theme, but have also included such exotic events as a series of poetry readings and the production of operas. Most of these events have been funded by grants to AAS from such funding agencies as the National Endowment for the Humanities, the Massachusetts Council on the Arts and Humanities, the Massachusetts Foundation on Humanities and Public Policy, and the Worcester Cultural Commission. Although these offerings have been wonderfully diverse, ranging from a lecture series on the social impact on New England of the American Revolution to one explicating the histories of food, drink, and sex in America's past, all are united by their origin in the Society's desire to interpret its collections and the kinds of scholarly research taking place there to audiences of lay people in Worcester and environs. [J.B.H.]

RESEARCH

Making scholarly research possible is what the American Antiquarian Society is all about. Mostly, such research is carried out by the thousands of scholars who travel from all over the world to work in the Society's library. The 'research' part of the department's mandate is intended both to provide the means by which its staff carries on research and to provide directed activities in which others take part. The Society's associate director for research and publication coordinates funded research projects within the Society. Such undertakings have included the Catalogue of American Engravings and the North American Imprints Program, both funded in part by the National Endowment for the Humanities. The former is a checklist of all engravings that were published in America through 1820 either as separates or as illustrations in books. The latter is an undertaking that will result in a computerized database describing all books, pamphlets, and broadsides printed in America through 1876. The earliest phase of NAIP carries the work through 1800 and cooperates with the Eighteenth Century Short Title Catalogue project organized by the British Library.

A recently established research program has educational overtones as well. It is the Program in the History of the Book in American Culture. AAS established the Program in 1983 in order to focus the Society's resources on promoting an emerging field of interdisciplinary inquiry. In doing so, AAS draws not only on its traditional resources as a center of bibliographical research and as a matchless repository of printed matter, but also on certain intellectual currents from abroad that look at the history of books in their full economic, social, and cultural context. In order to accomplish its goal of providing intellectual leadership in this field, the Program has sponsored conferences, publications, seminars, and research fellowships. A summer seminar in the history of the book, offering short-term, intensive training in methodologies and concepts, was initiated in 1985. The seminar has been successful in assembling a stimulating range of persons concerned with the field. A conference that included European scholars was held in 1984, and resulted in the publication of *Needs and Opportunities in the History of the Book: America, 1639–1876*. An earlier conference in 1980 resulted in the publication of *Printing and Society in Early America*, a collection of original essays, some of which have since become widely cited. The annual James Russell Wiggins Lecture in the History of the Book in American Culture, inaugurated in 1983, has brought forth important conceptual statements by leading scholars in different disciplines touching on the field. A thrice-yearly newsletter, *The Book*, serves as the chief means by which the Program communicates with its various constituencies and publishes substantive pieces as well as notes on research collections and on research in progress. A significant goal of the Program is the publication, in the early 1990s, of a multivolume, collaborative history of the book in American culture. [J.B.H.]

DEVELOPMENT AND PUBLIC RELATIONS

❧

The need to raise funds is a common theme running through the council reports from the founding of the American Antiquarian Society, although a department dedicated exclusively to fundraising was not established at the Society until 1968. By definition, an independent library depends on private support to survive and prosper, and Isaiah Thomas was eloquent in his appeals for funds, even prefacing his will with an admonition to his fellow citizens to contribute to institutions established for the public good. In 1820 Thomas provided the site, $2,000, and 150,000 bricks for the construction of the first Antiquarian Hall. His legacy, received in 1831, was given with the stipulation that fire-proof wings be added to the original structure.

The second library (1852) was constructed on land donated by Stephen Salisbury II, who served as president of the Society from 1854 to 1884. Salisbury contributed most of the funds for this building as well as for an addition built in 1878.

The present Antiquarian Hall, built in 1910, was funded by a bequest of $200,000 from Stephen Salisbury III, AAS president from 1887 to 1905. It was constructed on land that originally had been given to the Worcester Art Museum but which AAS exchanged for the Salisbury Mansion, which had been bequeathed to this institution.

Unlike earlier AAS construction projects, an addition made to Antiquarian Hall in 1971 was not funded by a contribution from a single donor or a few wealthy donors; rather, dozens of individuals, corporations, and foundations made it possible. Thus, the formal program of fundraising was set, a program that would become the pattern for future efforts made by the Society.

The first development officer, Elliott B. Knowlton, was appointed in July 1968, after the Council had voted to establish a capital campaign for endowment and construction needs. He served in this capacity until his retirement in 1976. Frank L. Harrington, Sr., was general chairman of the Development Committee. He was succeeded three years later by Howard B. Jefferson.

During that campaign, nearly $3,000,000 was raised for endowed positions. The Andrew W. Mellon Curatorship of Graphic Arts (1973), the Alden Porter Johnson Endowment for Publication, the Fred Harris Daniels Fund (1975), and the curatorship of newspapers were all established during that effort. Also during this period, an addition containing staff offices, study carrels, a manuscripts room, and acquisitions and cataloguing departments was built, at a cost of $1,500,000.

In 1968, the semiannual *News-Letter* of the Society first appeared. Intended to bring news of the activities of the Society to members and the scholarly community, it became the responsibility of the development office. [M.V.C.C.]

ANNUAL GIVING

According to the original bylaws, the day-to-day operations of the Society was to be paid for in part by donations of two dollars (later changed to six) per year per member. This system does not appear to have been effective, and, in treasurers' reports from 1830 on, there was no mention of dues or levies for members.

After the death of Thomas in 1831, funding for operating expenses was derived mainly from the income from his bequest, which had established a fund that was deemed adequate and ample for the Society to be self-sustaining. Although the income was modest, so were the expenses. In 1854, the Society operated on a budget of $1,300.

General operating expenses for the Society continued to be frugal into the twentieth century. The first formal appeal to members was not made until 1934, when Clarence Brigham, badly shaken by the deaths of two of the Society's staunchest members, Waldo Lincoln and Calvin Coolidge, upon whom he had depended to aid him in raising capital, decided that, without an adequate endowment, he must depend on annual gifts from members. This step, considered automatic by most institutions today, was undertaken reluctantly by Brigham. The first appeal outside the membership was not made until 1973. Today, contributions to the Society's annual fund are in the amount of approximately $200,000 from more than seven hundred donors. At present, separate annual appeals are made to members, friends, corporations, antiquarian booksellers, printers and publishers, and AAS readers.

In addition to its members, the Society receives effective and encouraging interest and support from the members of the Worcester Association for Mutual Aid in Detecting Thieves. The association was founded in 1795 to pursue thieves and burglars. With the establishment of a publicly supported police force, its original purpose became obsolete and the organization became a social club. When the last living member, Glendon Pomeroy, turned over the archives and assets of the association to the American Antiquarian Society, Linwood M. Erskine, Jr., recording secretary of the Society, urged AAS to reactivate the association as a support arm of AAS. The first meeting of the reconstituted WAMADT took place on June 1, 1978. There are now more than four hundred Thief Detectors who contribute annually to the Society. [M.V.C.C.]

ISAIAH THOMAS FUND

A capital campaign designed to increase the Society's endowment was formally announced at the annual meeting in 1982. Named in honor of Isaiah Thomas, the campaign set a goal of $8.7 million with $1,000,000 for acquisitions, $1,000,000 for conservation, $4,450,000 for personnel, and $1,500,000 for education. $750,000 was donated for the renovation of the Goddard-Daniels House and to provide for architectural plans for an addition to the present library building. President John Jeppson, 2nd, accepted general chairmanship of the campaign and Robert Cushman has led recent efforts. Named endowment funds have proliferated in recent years, twelve having been designated within the last five years, and have been added to the list of generous benefactions that go back to the original gift of Isaiah Thomas.

Dr. William Paine, vice-president of the AAS Council, addressing the members of the Society on October 23, 1815, said that 'the only check to the rapid progress of the Society is our poverty.' Libraries by their very nature must expand and grow. To do that, independent institutions, such as the American Antiquarian Society, depend upon the generosity and interest of individuals and corporations. [M.V.C.C.]

BUILDINGS AND GROUNDS

ANTIQUARIAN HALL AND OTHER PROPERTIES

Not long after the establishment of the American Antiquarian Society, its founder, Isaiah Thomas, began to plan a building in which its library could be safely held and used. By 1818 he had received preliminary plans from Peter Banner, the Boston architect, but Thomas played a large role in the final arrangement of the structure which was dedicated on August 24, 1820. This, the first Antiquarian Hall, stood at the corner of Belmont and Summer Streets, the present site of the Worcester Police Department headquarters (formerly the famed Valhalla Bar). It was one of the earliest buildings constructed in this country that was specifically designed as a library.

The second Antiquarian Hall was built in 1854 on the opposite side of Lincoln Square, next to Bulfinch's 1803 courthouse. By 1876 the collections had grown to such an extent that an addition was needed, the cost of which was borne by Stephen Salisbury, II. Designed by the local architect Stephen C. Earle, it nearly doubled the shelf space for books and served well until the demand for still more space required new arrangements after the turn of the century.

The building of the present Antiquarian Hall at the corner of Salisbury Street and Park Avenue had far-reaching effects upon the course of the Society's work. When president Stephen Salisbury III died in 1905, the Society received a bequest of $200,000 and a parcel of his land, obtainable from the trustees of the art museum, on which to build a new Antiquarian Hall. With the construction of this third building, the Council consciously altered the nature of the Society's mission. The collection of artifacts, known as the cabinet, the gathering of which had been part of Isaiah Thomas's original plan, was abandoned. Out went the stone ax heads, the pre-Columbian artifacts, the mummified

Indian maiden from Mammoth Cave, the gigantic plaster casts of Michaelangelo's Moses the Law-Giver, and the facade of the temple at Labna in Yucatan. AAS directed its energies toward developing its research library collections (which, at about 99,000 volumes, were very much stronger than its miscellaneous artifacts). So the third Antiquarian Hall was built as a research library, on a Palladian model reminiscent of the 1820 building.

Clarence S. Brigham, who arrived as librarian in 1908, and Waldo Lincoln, president of the Society from 1907 to 1927, were the architects of the Society's new programs. The new library was designed by the firm of Winslow, Bigelow, & Wadsworth, and its construction was overseen by R. Clipston Sturgis. The building was opened to the public in mid-1910. The rotunda, which serves as the reading room, was surrounded by alcoves and four rooms, three for offices and the fourth as the Council's meeting room. Upstairs, on the mezzanine level, the graphic arts and the manuscript collections were housed in two large rooms; other smaller rooms were used to display Thomas's first printing press and pieces of old printing equipment from the Worcester *Spy* office. Various pieces of furniture, some of it from John Hancock's elegant Beacon Hill residence, were scattered about in a forlorn sort of way.

The five-level bookstack was constructed in the best, fireproof manner of the day, with free-standing steel shelving running from cellar to ceiling, glass floors, and electrical wiring run in separate conduits. An advanced heating and ventilating system was also installed. Unfortunately, the large electric motors burned out a few years after installation and were not replaced. Two five-level additions were made to the bookstack in 1924 and in 1950. In 1970–71, an office addition was built and the rotunda area was substantially altered. The shelved alcoves were removed, which opened up the reading room, the Council Room (which had become the director's office) was made into an exhibition room, and the manuscript room was rebuilt to hold the Society's most precious books and to serve as the Council Room. The 1910 heating system was replaced, and a new air-conditioning system was installed. Additional stack areas were built for special collections and for manuscripts, with the inclusion of five carrels for visiting scholars. The addition, which is in large part underground, was designed by James Ford Clapp of the firm of Shepley, Bulfinch, Richardson, and Abbott of Boston. Antiquarian Hall was designated a national historic landmark in 1968 and is the anchoring structure of the Massachusetts Avenue Historic District. Now, in 1987, we are looking forward to the need for yet another addition to Antiquarian Hall, in order to provide more extensive and modern book stacks, a more efficient and secure reading room, expanded facilities for scholars in residence, and improved work spaces for our growing staff.

In 1981, AAS came into possession of the Goddard-Daniels House, located across Salisbury Street from Antiquarian Hall. It had been deeded with a life tenancy clause to the Society in 1970 by Eleanor Goddard Daniels, the daughter of the builders of the house and the widow of our councilor Harold F. Daniels. The house was constructed in 1905 and was extensively reconstructed in 1915 at the time of Eleanor's marriage. Following Mrs. Daniels's death, AAS made a number of renovations to fit it for institutional use, but its primary rooms were left essentially unaltered and are used for seminars, lectures, or for social functions. Secondary spaces were changed to accommodate office space for the staff of the research, publication, and education department and a small, modern kitchen. Bedrooms on the second story are made available to visiting scholars, and the music room, sans pipe organ, is now the fellows' lounge. The third floor was renovated and is now the living quarters of the Society's superintendent of buildings and grounds. The grounds are quite beautiful and are the site of a number of uncommonly found species of trees.

Upon the death of Mary Gage Rice, a lifelong friend of the Society whose family was associated with the Society since 1878, the Society received a bequest of nearly a half million dollars as well as Mrs. Rice's residence at Four Military Road, a few blocks from Antiquarian Hall. Since July 1978, it has served as the residence of the Society's director and is called Rice House.

In 1982, the Society purchased the property that makes up the house lot of Nine Regent Street which extends from Regent Street on the north to Park Avenue on the east. This parcel abuts the southern boundary of the land on which Antiquarian Hall sits, and control of this property is essential to the future growth of the Society. A substantial garage facing Park Avenue is used to store AAS equipment and the house is maintained as a rental property.

Some years ago, the small stable to a house that once stood as the corner of Salisbury and Montvale Road was converted into a cottage by Mr. Daniels. Because it so closely impinges on the Goddard-Daniels House, the Society purchased the cottage in 1983 and at present rents it to a tenant. [M.A.McC.]

Notes on Contributors

The following key identifies the contributors to this volume. The date within parentheses after each name indicates the first year of full-time employment at the Society. In the case of present staff members, positions at the time of publication are listed. In the case of individuals who have retired or resigned, the last position held at AAS is given, along with present residence or occupation.

C.A.A. Carolyn A. Allen (1971), acquisitions administrator

E.S.A. Eleanor S. Adams (1953), executive assistant to the director

K.A. Keith Arbour (1980), head of readers' services; resigned to pursue doctorate in history at the University of Michigan, 1987

R.L.A. Richard L. Anders (1968), cataloguer

F.E.B. Frederick E. Bauer, Jr. (1970), associate librarian; retired to Mirror Lake, N.H., 1984

G.B.B. Georgia B. Barnhill (1968), Andrew W. Mellon curator of graphic arts.

N.H.B. Nancy H. Burkett (1973), assistant librarian

P.A.B. Paula A. Biegay (1984), assistant cataloguer in the North American Imprints Program

R.C.B. Richard C. Baker (1982), chief book and paper conservator

J.D.C. Joanne D. Chaison (1981), head of readers' services

J.F.C. James F. Cuffe, Jr. (1980), cataloguer in the North American Imprints Program; joined U.S. Peace Corps, 1985

J.K.C. Jill Ker Conway, vice-president of the Society, visiting scholar at MIT, and a former president of Smith College

M.V.C.C. Mary V. C. Callahan (1976), development officer

A.N.D. Alan N. Degutis (1977), head of cataloguing services

R.C.F. Richard C. Fyffe (1981), senior cataloguer in the American Children's Books Cataloguing Project; became librarian of the Essex Institute, 1986

M.G.	Martha Gunnarson (1982), co-supervisor in the Newspaper Cataloguing Project; became a cataloguer at Worcester Polytechnic Institute, 1987
S.J.W.G.	Susan J. W. Gordon (1982), cataloguer in the American Children's Books Cataloguing Project
J.B.H.	John B. Hench (1973), associate director for research and publication
C.R.K.	Carol R. Kanis (1974), cataloguing specialist for authority
D.R.L.	Dennis R. Laurie (1978), assistant in newspapers and serials
M.E.L.	Marie E. Lamoureux (1976), genealogy specialist in readers' services
M.S.McA.	M. Sheila McAvey (1983), assistant editor
M.A.McC.	Marcus A. McCorison (1960), director and librarian
A.C.M.	Anne C. Moore (1982), cataloguer in the North American Imprints Program
P.A.M.	Pamela A. Meitzler (1983), cataloguer in the North American Imprints Program
D.N.O.	Doris N. O'Keefe (1979), senior cataloguer
J.M.P.	Joan M. Pingeton (1977), acquisitions assistant
B.T.S.	Barbara Trippel Simmons (1984), curator of manuscripts
J.A.T.	Joyce Ann Tracy (1973), curator of newspapers and serials
A.T.Z.	Audrey T. Zook (1973), assistant in serials

Index

175th
AMERICAN
ANTIQUARIAN
SOCIETY
1812 - 1987